Introduction to Environmental Ethics

A comprehensive introductory textbook that integrates environmental ethics with social and religious perspectives, this cross-disciplinary approach helps students understand the depth and breadth of environmental issues in a globally relevant way. Environmental problems are explained by ecological, social, ethical, and religious dimensions that inspire environmental conservation. It underlines the reasons why intrinsic values of nature, and those who inhabit it, must be taken into consideration when making decisions that affect them. This book is also intended for those interested in understanding the environmental debate, which is a critical component of environmental management.

Features

- Presents the historical review of the environmental movement.
- Introduces ethical reasoning in environmental problems.
- Explains environmental ethic approaches and practical implications.
- Discusses the role of religions in nature conservation and ecotheology.
- Includes exercises for reflection in each chapter to stimulate student engagement with environmental issues.

This great textbook gives readers new insight into caring for the environment. It is for undergraduate and graduate students taking courses in environmental studies, geography, sociology, anthropology, and philosophy, as well as professionals in environmental management and environmental policy, and for those interested in environmental issues and the ethics of nature conservation.

Introduction to Environmental Ethics

Caring for the Earth

Emilio Chuvieco and
María Á. Martín

CRC Press
Taylor & Francis Group
Boca Raton London New York

CRC Press is an imprint of the
Taylor & Francis Group, an **informa** business

Designed cover image: © Shutterstock

First edition published 2025
by CRC Press
2385 NW Executive Center Drive, Suite 320, Boca Raton FL 33431

and by CRC Press
4 Park Square, Milton Park, Abingdon, Oxon, OX14 4RN

CRC Press is an imprint of Taylor & Francis Group, LLC

ISBN: 9781032900490 (hbk)
ISBN: 9781032900544 (pbk)
ISBN: 9781003545972 (ebk)

DOI: 10.1201/9781003545972

Typeset in Palatino
by Newgen Publishing UK

Contents

Foreword

In an era of unprecedented environmental challenges, *Introduction to Environmental Ethics: Caring for the Earth* arrives as both a timely and transformative contribution to the discourse on humanity's relationship with the natural world. Emilio Chuvieco and María Á. Martín offer a profound exploration of environmental ethics dimensions of global change, weaving together insights from science, philosophy, and theology to illuminate the path forward in our collective responsibility to steward the planet.

This book goes beyond surface-level discussions of environmental crises. It delves into the roots of conservationism, the complex interplay between culture and nature, and the philosophical frameworks that shape our understanding of what it means to care for the Earth. The authors provide a holistic view, from tracing the historical evolution of environmental movements to examining the moral imperatives underpinning modern conservation efforts.

One of the book's most compelling aspects is its truly multidisciplinary approach, seamlessly blending ecological science with ethical and spiritual perspectives. By addressing both the tangible impacts of human activity and the intangible values that guide our decisions, Chuvieco and Martín challenge us to reconsider not only how we interact with the environment but also why it matters. Their call to integrate ethical considerations into environmental actions resonates deeply in an age where technological solutions and simple cost-benefit calculations often overshadow the need for a deeper value-driven paradigm shift including recognizing and valuing the priceless gifts of nature.

This work is a vital resource for educators, policymakers, and anyone committed to understanding and addressing the environmental challenges of our time. It invites readers to reflect, question, and ultimately act in ways that honor the intrinsic value of nature and its multifaceted connections to human well-being.

As you embark on this journey through the pages of *Introduction to Environmental Ethics*, I encourage you to approach it with curiosity, humility, and an openness to reimagining the ways we relate to our shared home. The wisdom contained within these chapters is not only a guide for better understanding the environmental crisis but also a blueprint for fostering a more sustainable and compassionate world for both humans and nature.

Michael Obersteiner
Director of the Environmental Change Institute
School of Geography and the Environment, University of Oxford

Preface

Beyond the different visions of life that coexist in our contemporary mentality, most people consider respect for nature as a value that belongs to a developed society. However, environmental concerns rarely have a direct and daily impact on people's lives, giving way to other issues considered more relevant, such as employment, public safety or access to social services, if they are not ignored out of indifference or selfish preferences linked to the satisfaction of whims. In practice, there are still few of us who believe that the conservation of nature is part of the common good and therefore above individual interests, not only because it benefits us all, but even more because it is the basis of human survival itself. In an eminently materialistic society, nature tends to be valued in economic terms, such as the benefits we receive – for example, scenic beauty that can be marketed – or in terms of investment constraints, either because of the resources that are conserved or because of the costs of using cleaner technologies. It is much less common to consider the intrinsic value of nature and those who inhabit it (plants, animals, other human beings), and consequently to take them into account in decisions that affect them, as something more than mere resources at our service.

Nature, understood as the set of animate and inanimate elements that make up the physical environment in which we move, is much more than a commodity. The rivers or mountains, plants or animals that share the planet Earth with us are also an inexhaustible source of teaching, a tremendous gift that delights our spirit, a mirror in which to better understand our human condition, which ultimately brings us closer to our fellow human beings and to our Creator. Throughout human history, nature has served not only to satisfy man's needs for food, shelter and refuge, but has also fulfilled his transcendent desires, from the search for beauty (poetry, painting, music, etc.) to the contemplation of that something divine that Creation manifests.

Many authors concerned about the environmental situation agree that the solution to the problems is not only technical; rather it requires a change in our current social model. This also implies a change in the values that guide this model, replacing materialistic consumerism, which identifies happiness with the mere accumulation of goods, with another approach that prioritizes more intangible elements, more in line with the spiritual nature of human beings. This change will be based on a greater appreciation of the natural environment, but also on a greater awareness of our ethical and moral commitment to the environment.

Various thinkers and religious leaders, from the Dalai Lama to Patriarch Bartolomew or Pope Francis, have emphasized that the ecological crisis is not just a technical problem, but rather a moral dilemma, that the root of the problems lies in our attitudes and values, or rather, in our lack of values.

Changing attitudes also means changing some of the things we do, the goods we buy, the food or energy we consume. It means changing our cosmological vision, reflecting on our role on the planet, our responsibilities as individuals, as a society, and as a species.

In these pages, we propose to present the different cosmological visions that underlie environmentalism, the roots of the growing interest in the environment, and some of the consequences that they generate. Before doing so, we will present some introductory concepts and some of the major environmental problems we face.

We are aware that many readers may feel a certain aversion to environmental movements, which are often associated with political or economic positions with which they disagree. We believe it is important to distinguish between the issue and the approach. The relevance of environmental problems requires that many people, from very different perspectives, work together to change our current relationship with the rest of nature. It is not essential that we agree on how to do this, nor, of course, on the consequences that these different approaches may have on other issues. Working together to solve a global problem requires joining forces with people of very different beliefs and convictions. Let us focus on what unites us and leave aside other issues that are largely peripheral to the environmental problem.

This text is the result of our academic activity and our personal concern. In the first aspect, it is only fair to thank the people who have accompanied us on this intellectual journey, especially through the summer courses organized by the Chair of Environmental Ethics of the University of Alcalá between 2013 and 2019. The comments and suggestions made by Josep Maria Mallarch, Mario Burgui and Alfredo Marcos on this text have been very valuable. We would also like to thank the students of the University of Alcalá and of the Rey Juan Carlos, whose questions and comments have often helped us to reflect on aspects that we ourselves had not considered.

We hope that this book will serve to deepen our understanding of the environmental situation of the planet, to learn more about the different philosophical and religious viewpoints that propose scenarios for change and, above all, to encourage our personal active stance in the face of the environmental challenges that lie ahead of us.

Emilio Chuvieco and
María Á. Martín

About the Authors

Emilio Chuvieco is a Professor of Geography at the University of Alcalá, where he leads the research group in Environmental Remote Sensing and the Chair of Environmental Ethics. He has taught postgraduate courses in 16 countries. He has been a visiting researcher at the Canadian Remote Sensing Center and the Universities of Berkeley, Cambridge, Santa Barbara, Maryland and Oxford. He has coordinated 36 competitive research projects and 23 contracts. He has supervised 44 doctoral theses. He is co-author of more than 430 articles and book chapters. His main line of work focuses on Environmental Applications of Earth Observation Satellites (Forest Fires, Global Change, Deforestation), although in recent years he has also worked on theological reasons for environmental conservation. He has received the Excellence in Research award from the University of Alcalá and the Rei Jaume I award for Environmental Protection. He recently completed a DPHIL programme in Ecotheology at the University of Oxford.

María Á. Martín is currently a professor of Environmental Impact Assessment at the Universidad Rey Juan Carlos of Madrid and coordinator of the Environmental Ethics course of the Official Master in Bioethics at the same University. She holds a PhD in Biological Sciences from the Technical University of Madrid (UPM) and a Master of Environmental Management (MEM) from Yale School of Environment. She has been a visiting researcher at the Universities of Chile, Syracuse and Harvard. Her main line of research is ecologically based physical planning and in recent years she has worked on environmental governance and environmental conflict resolution.

Acronyms and Abbreviations

ARC	Alliance of Religions and Conservation
CF	carbon footprint
CH_4	methane
CO	carbon monoxide
CO_2	carbon dioxide
COP	Conference of Parties
GHG	greenhouse gases
IPCC	Intergovernmental Panel on Climate Change
LFA	life cycle assessment
LS	Laudatio Si'
NGO	non-government organizations
N_2O	nitrous oxide
NOAA	National Oceanic and Atmospheric Administration, climate center of the USA
SDG	Sustainable Development Goals
UN	United Nations
UNFCC	United Nations Framework for Climate Change

1

Defining the Terms

What Is Natural?

A few years ago, the Spanish press reported that an animal rights group had broken into a fur production-oriented farm and liberated thousands of American minks causing severe economic loss for these farms. At first glance, the news could be interpreted as the work of a radical environmentalist group, but the environmentalist NGOs immediately denounced this act, both because they did not share the violent methods used, and because it implied the release of an exotic species that would have a very negative impact on the local fauna and flora, as the American mink is larger and more voracious than the European one. For many lay people, animal and environmental activists are aligned, but this example shows that both groups have different goals and backgrounds, which do not necessarily agree. In fact, on many occasions, such as this one, they are in open disagreement, since the protection of natural ecosystems requires avoiding exotic species, which animal right groups consider as valuable as the endemic ones.

This example shows that environmental protection has very little to do with the application to animals of concepts (freedom, rights) that have been traditionally considered exclusively applied to humans. Along the same line, the environmental debate of accepting or not a fur farm is not so much whether it is ethically appropriate to farm animals for their skins, but rather if it is ecologically harmful. Assuming that fur trading is an ethical activity, few will doubt that it is better to use the skins or meat of animals raised for that purpose than to hunt them in their natural environment.

Another issue is the growing appreciation of animals -and the support of their preservation- within the contemporary sensibility, commonly more related to the aesthetic features of the animal than to their ecosystem role. In this regard, the values of an environmentalist would differ from those of lay people. For instance, native insects that are generally not highly regarded in terms of conservation status may be more relevant than picturesque but sometimes exotic mammals, such as deer or bear, which may have been introduced from other regions, because those little insects, even though

DOI: 10.1201/9781003545972-1

unpopular may be carrying out a fundamental role in a particular ecosystem and therefore are essential to its balance. In a similar way, bees are much less attractive than pets, but they play a key role in pollination, which in turn is key to the reproduction of many plants, while pets have a social role that only benefits human beings.

In short, these ideas make us think about what exactly it is that we want to conserve when we aim to conserve nature. In a survey conducted some years ago on the Internet about what nature was, many respondents emphasized its non-human character, as if human beings were some kind of *alien* who had fallen on the planet from outer space. Certainly, in common parlance a landscape is considered natural when it shows no signs of human transformation. Natural, in this context, is synonymous with pristine, native or wild, as opposed to an artificial landscape where human imprints predominate, whether in the form of infrastructure (roads, railroads), construction (buildings), production (agriculture, forest plantations) or ornamental design (gardens, playgrounds).

From the most obvious human interventions to others that are less evident, such as those caused by centuries of reforestation, fires or introduced plants and animals, the range of human coexistence with nature is very wide. The extent to which we consider something natural as a 'space without human activity', largely defines the area which we would eventually have an interest in preserving.

If we use the term 'nature' in a very restrictive sense, the surface of the planet that we could consider natural is very small. In fact, most of the Earth bears the imprint of human activity in one way or another. Although ecosystems have certainly undergone multiple changes without human intervention, those due to the direct action of our species are unparalleled to those caused by other species. We have modified the original vegetation cover to grow crops, feed our livestock, build homes or generate energy. In large parts of Europe, the Near and Far East, Central America and more isolated parts of South America and Africa, signs of human activity date back more than four millennia, so natural landscapes – if we use the above meaning – are exceptional. And yet we have many examples of landscapes that are valued as natural, and even protected as such, that are far from being free of human intervention.

Since what is strictly natural is so limited, we normally adopt a less restrictive definition when we refer to nature conservation, including also areas with little presence of artificial features, or we even grant protection to cultural landscapes, which are precisely the result of man's historical activity, such as the agricultural terraces so traditional in the Mediterranean or in the East Asian mountains. In short, we can consider environmental conservation in a broader framework, where the human being is part of the environment as another element, but without disturbing its harmony, without being such a protagonist that we only observe human features in the territory.

This stricter sense of what is natural can lead to more radical attitudes in the conservation of nature, as we will see later, which see humans as the main enemy of conservation. If natural is only that alien to humans, then the only way to preserve nature is to prevent humans from interacting with it. In short, according to this approach, the best and purest nature is the Earth where we are not, where we are only visitors. The place where we are is the place where nature is not. This form of primitivism in this century resulted in what some call the cult of regions in a natural state (Lorentzen 2001, p. 39).

Although some authors continue to believe that the best way to conserve nature is to isolate it from human beings,[1] fortunately in our opinion this reductionist view of nature is now less extended than in a few decades ago, and there is a growing tendency to consider as natural also territories that have or have had human influence, particularly from native cultures. As a result of this greater appreciation of the human presence in nature, the role played by indigenous peoples in the conservation of the territories in which they have lived for centuries is increasingly valued. In past decades, the most radical conservationists advised the expulsion of these indigenous peoples. Today, it is considered that these communities have maintained a balance with the environment that guarantees the functioning of ecosystems, since they are part of those ecosystems. In other words, plant biodiversity does not necessarily conflict with cultural biodiversity, but rather in many cases supports it.

In this discussion about what nature is and is not -and, therefore, about what we should preserve – it seems to us very relevant to recover its original philosophical meaning. The classics understood nature as the essence of things, in short what makes them a certain thing and not something else: a forest, a stream or a tiger have their own essence, also a purpose that they fulfill in the ecosystem where they live, precisely because they are what they are and not something else. If the tiger were not a predator of antelopes, and these had no other animal to eat them, they would grow so much that they would not have enough vegetation to survive, in addition to greatly degrading the existing vegetation in the process. A tiger must then do what a tiger is supposed to do (in this case, eat an antelope) and not do other things that its '*tigerish*' nature is not supposed to do, if we can say so, such as doing pirouettes in a circus. In this sense, the conservation of nature would be the preservation of the purposes for which that animal is 'designed', the designer being the evolution of the species itself (randomly or specifically willed by God, depending on one's interpretation of the ultimate drivers of the evolutionary process). In both cases, we are assuming that the natural state is the most perfectly adapted to the final role that each component plays in the ecosystem.

In short, to conserve nature would be to keep it as it should be (in its deepest sense), respecting the physical or biological laws that have originated it. This approach should also include human beings, since we are part of the very environment we are trying to preserve. In this sense, it seems important to us to emphasize the close relationship that should exist between ecoethics and bioethics, that is, between reflection on the principles and actions concerning people and the natural environment, not only because they are part of it, but also because they have a nature of their own (human nature), that a convinced naturalist should also maintain. Thus, to preserve nature is also to keep the essence of the human beings, what we are by nature, in the face of the invasion of technology that will supposedly perfect us (as the transhumanist movement proposes), as falsely as it does the natural environment itself. It is not a matter of falling into a technophobia, as Luddites did in the 19th century, but of using technology to improve our lives, rather than to replace them (Marcos and Pérez Marcos 2018), as we will have the opportunity to comment in more detail later (see Human Ecology, Chapter 7).

Environment, Ecology, and Environmentalism

Since protecting the environment has become a generally accepted value in contemporary societies, that something is 'natural' is generally now considered as having superior value to its artificial counterpart. In this regard, the word ecology/ecological has come to qualify any element or action that is close to 'natural' conditions or, at least, little altered, and therefore of better quality that a similar product not having that qualification. For example, the term ecological livestock or agriculture is used to refer to meat or vegetable production with minimal use of artificial elements, and therefore healthier. By extension, people speak of ecological light bulbs, paper or cars when their consumption is lower or their manufacturing process has less (or none) negative impacts on the environment, and therefore these products should be more attractive for consumers. Whether this assumed environmental friendliness is real or pretended is difficult to verify, because traceability of carbon or water footprint of ordinarily consumed products is very unusual (see Changing Lifestyles, Chapter 7 for further discussion on this issue).

In recent years, the label 'ecological' has been replaced in many spheres by 'sustainable', which is even a more confusing term. Something is supposedly sustainable when it meets certain environmental conditions, which makes the institution, company or product to which the label is applied more environmental-friendly. Due to so much repetition, the adjective is now far from its origin,[2] since anything that tends to last over time can be considered sustainable (a certain business, for example), although in fact it had little to do with the environment. Arms manufacturing is certainly a sustainable activity (this business has been active for many centuries), but it is far from being environmental-friendly.

Someone having concern for the environment is often identified with being part of an environmental NGOs, but it should be clarified that interest in nature was neither invented by these groups nor is it exclusive to them. In more academic settings, the term ecology (*oikos-logos*, 'knowledge about our habitat') indicates the study of the relationships of living beings with each other and with the environment around them (climate, water, soils, vegetation, etc.), and ecologist is someone who dedicates his-her professional interest to the study of ecology. This science was first proposed by the Prussian biologist and philosopher Ernst Haeckel in 1866, although it was the Danish botanist Eugen Warming who wrote the first manual and taught the first course on the subject in 1895 and is therefore considered the founder of ecology. In 1927 Charles Elton pointed out that *'ecology is a new name for something very old'*, referring mainly to Natural History. Today this science is fully recognized as such in most universities, usually as a branch of Biology.

Ecosystems are the object of ecology. The concept of ecosystem was introduced by Tansley in 1935. It indicates a set of biological (organisms, substrate) and inorganic (energy, geological stratum) components that coexist in a place maintaining reciprocal relationships and where 'inputs' and 'outputs' of matter and energy can be quantified. In an ecosystem, energy usually flows from the sun, passes through the biotic community and its food web, and leaves the ecosystem in the form of heat, organic matter and organisms produced in the ecosystem. From the ecologist's point of view, ecosystems are generally the basic units of conservation. Rather than a concrete unit, the ecosystem is a level of organization from which the study of the structure and function of any part of nature is approached. A tree, a pond, a city, a continent, even the biosphere, would be ecosystems – at different scales – within which can be distinguished a set of ecosystems subordinated to one another and related by more or less intense flows of matter and energy. Traditionally, ecology follows two main lines of study. The first deals with the relationships between species and their physicochemical environment and the second encompasses population ecology (Cain et al. 2014; Mackenzie et al. 2001).

It is important not to confuse the term environmental scientist with environmental activist, which is also the term *ecologist* in some languages. The former is a scientist who studies the relationships between the organisms that make up a given territory, who can have a background in ecology or in other disciplines related to the study of natural processes: agronomy, forestry, geography, sociology, economy, biology, geology, etc. An 'ecologist' in the second sense, indicates a person who actively participates in an organization whose objectives are oriented towards environmental conservation. The range of objectives and means of action of these organizations is very varied. In general, they carry out a very commendable work of environmental awareness, although some organizations may use lobbying methods or support political agendas in which environmental issues are mixed with other social themes, which can create reluctance even among people who are generally in favor of conservation. It should also be noted that environmental movements do not have a uniform approach, but may adopt conflicting strategies, for example, in their active or non-active role in political parties, or in the use of peaceful resistance or active violence in their activities (see Conservationist Social Movements, Chapter 3).

Book Reading Guide

Within the scope of this introductory chapter, it is worth placing the ecological debate in a broader context that attempts to consider the many different aspects and values to which this issue is related.

Although it may seem obvious, the first question to be clarified when dealing with the ecological problem, as with any other, is to define with rigor exactly what its magnitude is, how important it is in relation to other problems that may also require considerable social or economic effort. For many authors, environmental problems, and specifically climate change, are the most important challenges facing humanity today, since they threaten the very existence of the human species on Earth. This is the opinion of the authors of some of the ethics that we will discuss in later chapters: they have obviously devoted their intellectual activity to this subject because they believe it to be crucial. If the facts did not support this approach, or, to put it another way, if the magnitude of the ecological problem were in fact minor or even marginal, the reasonable conclusion would be not to take measures that would be a distraction from solving other problems of greater relevance. From the point of view of the reader of this work, if the arguments showing the importance of the problem (see Chapter 2) conclude that it is a minor issue, it would be best to close the book and devote the time to other more useful reading.

If we succeed in convincing the skeptical reader, or confirming the one who is already convinced, of the seriousness and centrality of the current ecological problem, we should then identify who is responsible for the problem. There are not many alternatives, since environmental degradation can only be a consequence of natural or human factors. Which of the two has greater weight in such pressing problems as desertification, biodiversity loss or climate change? If we conclude that natural factors are the protagonists, this book should have a biological, geological or atmospheric approach to help us understand the problem and try to minimize its negative impacts or rather introduce political guidelines that would help us to be better prepared for its consequences. However, if the problem is of human origin, the solution should have an ethical component, since it would require the adoption of measures that would lead not only to the reduction of environmental damages, but also to be better adapted to their future consequences. In this book we will focus on ecoethical issues, thus we are implicitly assuming that human beings are the main ones responsible for environmental problems and, consequently, that their solution implies a change on the way we deal with nature. This does not mean that only human factors are leading environmental changes: natural factors also play an important role, sometimes emphasizing human triggered issues, sometimes alleviating them.

Having confirmed the importance of the human factor and the need to make significant changes in the way we treat nature, the next logical step is to reflect on the main reasons underlying this new attitude. One of the most obvious is that nature is the place where we live, our 'common home', to use the expression popularized by Pope Francis (2015). Deteriorating nature means damaging the place we live in, where we breathe, move and feed ourselves. There are also ethical or moral reasons for caring for the Earth. We will cover all these arguments in Chapter 4, with particular emphasis on ethical and religious arguments (Chapters 5 and 6, respectively) for environmental concern and action.

Once the reasons for supporting environmental concerns and actions are established, we should consider how to do so more effectively. One's environmental response will be largely determined by the reasons that motivate it. If the arguments are purely utilitarian, the best response would be to do nothing ('business as usual') or to minimize change, leaving the problem to future generations in an exercise of generational selfishness. Others may see industrialization as the cause of the current ecological crisis, and advocate the use of low technology and a return to simpler ways of life. Others may take the opposite view, believing in the power of technology, convinced that all problems can be solved by human ingenuity alone, and that we should therefore focus on investing in technological development. Yet another group may believe that humans are the cancer on the planet and would favor reducing the human population, either through incentives or coercion. In our opinion, the solution to the environmental problem is neither technological, although technology is a great ally of change, nor demographic, since the promotion of antinatalist measures has caused much human suffering, nor a return to an idealistic pre-industrial society, because it would be utopian (we cannot turn back the clock of history) and because primitive societies have also acted unwisely regarding the environment (see *Indigenist* Response, Chapter 7).

The necessary change would involve confronting our relationship with the environment in moral terms, either as a consequence of an ethical responsibility toward other living beings (human or not), or as a consequence of a religious commitment, where the role of the human being in the whole of Creation is appreciated. This will lead to a change in the consideration and contemplation of the environment, in how we relate to the elements that make up, in how we conceive and use them. In short, it seems to us that it is necessary to rethink our values towards a more integral conception of our relationship with the environment. This will mean, among other things, rethinking our ways of life, analyzing to what extent they affect the environment and affect us as human beings, and allow us to ensure our habitable future on this precious planet.

All these issues will be discussed in the following chapters of this book, before introducing a brief description of current environmental problems (Chapter 2) and a historical introduction to the evolution of conservationism (Chapter 3).

Exercise for Reflection

Environmental psychologists have characterized different human reactions towards environmental problems, identifying also various drivers that explain why people react in a very different way to the same observations. One of the most recurrent barriers for environmental action is the perception that our actions are irrelevant to solving the problems. This is particularly true when referring to global issues, such as Climate Change, in which an 'eco-fatality' ('nothing of what I do is relevant') attitude may prevent further action. Objectively speaking, our behavior has little repercussion in solving global emissions, for instance, but it is also true that world trends are the aggregation of personal values (in fact, several authors have estimated that more than 70% of national emissions come from personal consumption), but more important than that is the fact that one's concerns need to be translated into daily practices. In environmental problems, as well as in many other issues, personal coherency matters.

The exercise that we propose on these pages aims to reflect on the impacts of personal behavior. We illustrated this principle with the meditated reading of 'The Man Who Planted Trees' a very brief but beautiful story written by the French author Jean Giono in 1953. The short tale has been published by several companies, but Giono wanted to be distributed freely, so there are many public versions available, and it should not be difficult to access it. Readers may also watch an animated short film based on this story, who won an Academy Award in 1988. In summary, the tale describes the long and single-handed effort of Elzéard Bouffier, a shepherd living in a barren valley near Provence, who dedicated his whole life to restore the area by planting oak acorns. This constant and selfless attitude completely changed the landscape where he lived from a barren ecosystem to a lush one.
Questions for reflection after reading:

As you read the story, do you think Elzéard was a real character or a fictional character? Which would you prefer? Why?

Do you think Elzéard's attitude is reasonable, realistic, pragmatic? Is it possible to achieve such effects? Is it worth it?

What environmental problems you face in your daily life depend on your particular attitude and behavior?

What would change if you adopted a similar attitude as Elzéard to solve them? What effort would it take?

Notes

1 This approach is behind the concept of integral reserves, areas without any human intervention, which have implied that in some cases native populations have been expulsed from certain newly declared protected areas. The proposal of 'half-Earth' made by the American biologist Edward Wilson (2016) shared also this focus, arguing that in order to preserve the planet's biodiversity it was essential that half of the Earth be free of human intervention of any kind.

2 The word sustainable development, indicating development compatible with environmental conservation, was proposed in 1972, as we will comment in Chapter 3.

2

Environmental Problems: Science Beyond Ideologies

Don't Let the Truth Spoil a Good Headline

The tendency to exaggerate or repeat commonplaces to reinforce a point of view is quite common to many ideologies. The degree of indulgence we have with such deviation from reality will depend on its magnitude, its implications and our own values. It seems reasonable to be wary of oversimplified representations of very complex problems. It is true that the more serious the conflicts are perceived to be, the easier it will be to make decisions that require social and economic sacrifices, but this cannot be done by ignoring aspects of reality that are key to understanding it. Introducing abusive simplifications, mixing observations with hypotheses, punctual data with trends, local with global phenomena, interpolations with extrapolations, sometimes quite gratuitously, does not help in the long run to consolidate far-reaching ethical decisions, which should be based on the best science available and the pursuit of the common good, respecting – whenever possible- personal freedom. Even if the noble purpose is to move a political or citizen action that is considered essential, artificially forcing scientific results, despising opposite alternatives will always be, in the long run, counterproductive, particularly when the promised impact is ultimately proven to be false, since it predisposes public opinion in a very negative way for future actions.[1]

In any case, the mission of scientists is to provide the most reliable information available to identify a problem, including its causes and extent. If possible, scientists should also be involved in disseminating sound information of that problem, educating public media to present it in a rigorous way, avoiding exaggerations or bias. In the case of the hunger situation, regardless of whether the number of people affected is large or not, society should eradicate this problem, as we have a moral obligation to act even if only one person dies for lack of food that we have in abundance in other countries.

In the field of environmental protection, there is no doubt that decisions affecting our way of life on the planet are serious and necessary. However,

 DOI: 10.1201/9781003545972-2

the real state of degradation of the Earth does not always correspond to the apocalyptic vision that appears in some media or in the writings of environmental thinkers, perhaps more concerned with arguments and less with empirical data, resulting in an oversimplification of reality. We could give many examples of these exaggerations:

One of the most widely commented on cases of this trend was the report presented to the Club of Rome by the Massachusetts Institute of Technology, entitled '*The Limits to Growth*' (Meadows et al. 1972). It became a worldwide *best-seller*, with more than four million copies sold, and had an enormous impact on political opinion and actions in the following years. Based on available data, and a series of assumptions about growth and known reserves, the report inferred that by the end of the 1990s most of the planet's energy and raw material sources would have been exhausted, and that almost one fifth of the population would have succumbed to the foreseeable food crisis. Fortunately for us, most of these calculations and predictions have proved to be completely wrong, as was anticipated by several economists shortly after the report was published (Clark 1977).

In the book *The Dream of the Earth* (Berry 1988), page 9 states: 'By the year 2000, between 0.5 and 1 million species will have become extinct on the planet'. He does not cite sources. On two other occasions in the same book, p. 39 and p. 207, he indicates that about ten thousand species become extinct annually. In this case, the source, Harvard University biologist Edward O. Wilson, is cited. Since Berry wrote his book in 1988, at that annual rate some 120,000 species would have been lost by the year 2000, a far cry from the million species he indicates in the first chapter. It is very complicated to calculate the number of species that are really going extinct, since in fact there is a large proportion (some biologists estimate more than 80%) that we do not even know yet, although among the documented species it is estimated that one fifth have a significant risk of going extinct in the next hundred years (Balmford et al. 2005), for other authors it could be as high as 50%.

Water conflicts also often appear among the most conclusive environmental issues, although this is not always done with the rigor that the available data provides. For example, in 1993 Joye Starr stated at the Global Water Summit that 'nations like Israel and Jordan have only ten to fifteen years before agriculture and, ultimately, their food security is seriously compromised' (quoted by Vesilind 1993). Thirty years later, it does not appear that these countries have any particular food supply problems, but rather other political issues that are much more relevant. There is no doubt that water is one of the basic elements of life, and access to it must be ensured as a basic human right, so it is essential to be especially cautious when estimating future resources. In this regard, studies based on climate change models are not very conclusive, since precipitation patterns are much more difficult to predict than those of temperature (Oki and Kanae 2006). Recent studies emphasize the importance of food trade, which also involves moving large quantities of water (the water needed to produce it), and improving purification, reuse and irrigation systems (Petersson et al. 2021).

In 1972, Michel Bosquet predicted that world reserves of Mercury would last 13 years, lead reserves 15 years, gold reserves 17 years, zinc reserves 18 years, silver reserves 20 years and copper reserves 40 years (ending in 2012). None of these predictions have been confirmed. In fact, copper reserves increased since then with an estimated availability in 2017 (following USGS estimations) of another 40 years. On the other hand, it is not so much a matter of how much will copper or any other mineral be needed in the future with our current technology, but also what technological changes would affect the use of that particular resource. Many years ago, Julien Simon (1981), in his controversy with Paul Ehrlich, emphasized this issue with a very simple example. The amount of copper needed to establish communication between Europe and America was in the range of millions of tons when using submarine cables, while now using modern communication satellites only a few kilos are required, or even none as copper has been replaced by fiber optic. This implies a practical expansion of a particular resource, when it is related to the function it exercises.

The Canary in the Coal Mine

Logically, we do not want to conclude from these comments that environmental problems are invented or exaggerated by conservation groups. The examples cited above show that there are uncertainties when predicting future trends, and that those uncertainties are not well understood by non-scientists using science results to back their own ideas. It is very important to recall that uncertainty in science is not ignorance: it is just imprecision, difficulty to establish an estimation with high confidence. Uncertainty may come from the imprecision of measurement instruments or from modeling approaches that are not yet able to handle all relevant variables. The more we know a process, the more likely we are to estimate and eventually predict its evolution. For better modeling an event we need precise observations, which sometimes are not yet available and therefore we should rely on rougher estimates. For instance, the evolution of temperature and precipitation in historical periods relies on dendrochronology, which makes it possible to guess which thermo-humidity conditions occurred during the life of a long-standing tree or the frequency of fires in a particular region (Swetnam 1993). Obviously, these estimations are less precise (have higher uncertainty) than those derived from current weather stations but at least they give an estimation of past trends well before thermo-pluviometer sensors were available.

The more precise instruments are available, the lower the uncertainties associated with estimations. To honor those uncertainties, is important to be prudent to establish sound conclusions when uncertainties are high, but still decisions can be taken with available information: the lack of high precision does not justify the absence of actions. When the potential risk is very high, even a low probability based on existing facts can advise an intervention. This is the main basis of the prudence principle included in the World Charter for Nature of the UN:

> Activities which are likely to pose a significant risk to nature shall be preceded by an exhaustive examination; their proponents shall demonstrate that expected benefits outweigh potential damage to nature, and where potential adverse effects are not fully understood, the activities should not proceed.
>
> (United Nations, World Charter for Nature, Resolution 37/7, 1982, 11.b)

In environmental issues, especially those that are strongly influenced by temporal dynamics, it is key to see significant trends and timely indicators. It is well known that several decades ago coal miners used to bring a canary to the mine, accompanying their excavation work to alert them of dangerous emissions of firedamp gas. When the miners saw the canary fade away, it was a sign that the concentration of gas was beginning to become dangerous, because these birds are much more sensitive to these fumes than human beings. The increased sensitivity of these birds was essential to save the lives of the miners themselves in the face of rising gas concentrations. When this or that natural species has disappeared from the ecosystem it inhabited, it is an indicator that this ecosystem (its climate, its vegetation or its soils) is undergoing significant changes, even if our observation instruments are not yet capable of detecting this change. Thus, the disappearance of a species, even a minor one, is a clear testimony that the conditions of the environment are different, all the more different from its natural level the more species disappear.

The image of the canary in the mine can be extended to many environmental problems, but perhaps one of the most appropriate is that of climate change, which is undoubtedly the most controversial environmental issue today, both because of the magnitude of its effects and the global scope of its implications. A few years ago, skeptics of the problem claimed that warming trends were small and not always significant. Now there are few who hold that view, given the wide variety of evidences that point in the same direction, supported by the large number of data series currently available. The problem is not only the magnitude of the change but also its rapidity and extension. Climate involves very long-term trends (30–40 years), to which the human scale is not very sensitive. Waiting to make decisions until such a change becomes evident to us is as unwise as waiting for the canary (and perhaps even a miner) to die before leaving the mine. Considering the enormous inertia of the climate system, we can affirm that, when climate change is recognizable by all – or, in other words, when even the laypersons have a direct experience that the climate has changed–, it will be too late to make decisions because the change will already be irreversible.

Some people, perhaps driven by a misunderstood humility, claim that humans are insignificant in affecting the Earth's environmental system, which is controlled by physical laws that considerably exceed our ability to alter them. Certainly, our planet has overwhelming magnitudes, but it is also important to underline that it shows an extremely fragile equilibrium. Several authors have stressed the uniqueness of our planet as a source of life (Ward and Brownlee 2000), the result of an extraordinary confluence of 'convergences'. Our planet is not just one among the hundreds of millions that may exist in the immensity of the galaxies, but has some very extraordinary characteristics that seem to point in an effective direction to harbor life:

- The Earth has an ideal position in a medium galaxy (which allows the formation of metals, due to the suitable cosmic radiation).
- It is part of a planetary system of a medium-sized star, the Sun, which is neither too hot (which would lead it to emit very energetic radiation) nor too cold (which would not allow the evolution of life).
- It occupies a suitable place in the solar system, not too far from the Sun, which would make it too cold, nor too close, which would make it unbearably hot, in either case making it impossible to find liquid water.
- It has other 'nearby' planets large enough (Jupiter) to block most of the asteroids that would otherwise fall on Earth (those that did fall have catastrophic effects, drastically altering the planet's conditions).
- It is a planet of an ideal size, which allows it to have an atmosphere capable of regulating temperature, and a geological composition that allows it to have moving tectonic plates on a viscous surface, the origin of the relief variations we know, and which also generates magnetic currents that protect us from high-energy solar radiation.
- The Earth has an optimal tilt on its axis of rotation, which explains the existence of climatic seasons, key to the diversity of species we have.
- Finally, the Earth has a satellite of the right size to stabilize gravity and regulate the Earth's rotation speed, as well as explaining the movement of the tides, which also affect the movement of the continents.

In short, the very low probability of all these factors occurring in a single place has led various authors to enunciate the so-called 'anthropic principle', according to which our planet has followed a convergent evolution that would have been 'guided' towards the appearance of the human being. Science will not affirm whether this evolution is guided by a divine design or not, since this is a theological question that science itself cannot elucidate, but it will at least point out that the coincidence of so many factors is extraordinarily anomalous.

Considering all factors of this 'Earth uniqueness', one can realize that the environmental balance of the planet moves within very narrow margins. A small deviation in any of the above factors would have simply meant that life, as we know it today, would not exist. The basic question is whether humans are capable of altering this balance, with dire consequences for the system as a whole and, ultimately, for us as a species.

Some authors believe that our capacity to alter the territory is very small compared to the vastness of the Earth. But these authors overlook the immense advantage provided by human ingenuity, which allows us to adapt to any environment, using different techniques to overcome our biophysical limitations. It is enough to look at satellite images to realize that this global transformation of the planet's natural conditions has already occurred, especially through the enormous change of land cover that we have brought about. From the introduction of agriculture in the Neolithic period to the tropical deforestation of recent decades, the transformation of the Earth's surface caused by our economic activity can be rigorously described as planetary. It is estimated that more than 50% of the world's natural forests have been converted to pasture or crops. The conversion of forests to crops implies an alteration of the CO_2 balance in the atmosphere (as plants absorb CO_2 in photosynthesis to produce carbohydrates and store carbon in the form of biomass), an alteration of the hydrological cycle (as plants capture, retain and transpire water), a modification of soil characteristics (erosion, loss of organic matter) and a loss of biodiversity (forests are 'sanctuaries of life'). The conversion of forests to cropland or pasture is evident in many parts of the world, but in recent decades it has been particularly pronounced in tropical countries (Figure 2.1). Recent estimates based on satellite imagery indicate that in the last three decades a total of 218 million hectares of forest area have been lost in these tropical areas and another 165 million hectares have been degraded, which represents 10% of the total area occupied by these forests (Vancutsem et al. 2021), the planet's most important storehouses of biodiversity. In temperate forest the trend is contrary, with a general trend to increase forested areas as a result of rural abandonment or national forest afforestation policies.

Other human impacts on the environment refer to the deterioration of the oceans (acidification and overfishing), the intensification of agriculture (soil degradation and contamination), desertification (changes in rainfall patterns),

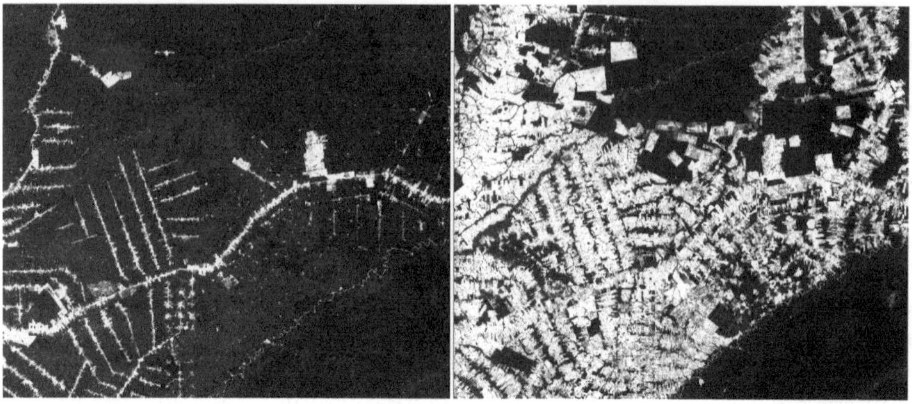

FIGURE 2.1
Landsat satellite images acquired in 1990 (left) and 2024 (right) over the State of Amazonas (Brazil). The transition from forest (dark) to agricultural (light) areas is evident. (Landsat USGS Glovis archive.)

and the degradation of landscapes due to the impact of mining, urbanization and transport infrastructure. The latter are more spatially localized, but have greater permanence over time, since they involve almost irreversible changes in natural conditions. The extent or urban areas shows clearly the global impact of human occupation (Figure 2.2).

In short, human impacts on the Earth are increasingly extensive and pervasive. This has led some authors to consider that we have in fact entered a new geological era, which is called the Anthropocene, because it is primarily characterized by human activity. The term was coined by the Dutch scientist Paul Crutzen (2002), who won the Nobel Prize in Chemistry in 1995 as one of the discoverers of the mechanisms that explain the depletion of the ozone layer.

Context and Trends

As mentioned above, environmental problems are dynamic, changing in response to natural and human factors. Tracking these trends is key to understanding the processes and predicting the consequences. In the case of climate change, this analysis is particularly evident, since it is precisely a matter of observing whether the observed trends are significantly different from the previously available time series.

In statistics, the word significant has a precise meaning: it is a matter of checking whether the observed data are within or outside a certain threshold

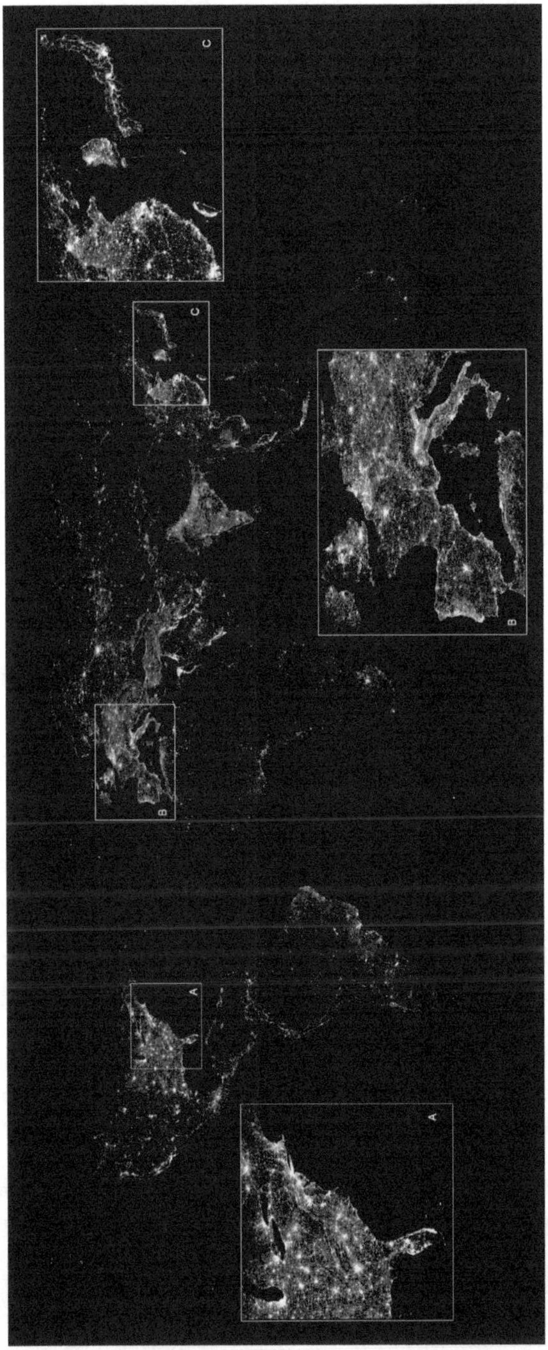

FIGURE 2.2
Satellite night image based on data acquired by the VIIRS sensor. Three highly urbanized regions are shown in the enlarged windows. Original data processed by the Earth Observation Group, Payne Institute for Public Policy, Colorado School of Mines. This data is distributed under CC-BY-SA license.

that would be expected by pure chance (randomness). In other words, a trend would be significantly different from a reference period when the new values exceed (above or below, depending on the trend) a threshold value that could occur simply as a consequence of the natural variability of the system. For example, if we state that the average summer temperature at a given weather station is 26 ± 1°C with 95% probability, we are indicating that the average temperature of a long time series is 26° and that in 95% of the years the average temperature has varied between 25 and 27°. If in that season we find a year with an average temperature of 30°, we will consider that year as anomalously hotter than expected. However, this summer value does not indicate a change in the trend, but rather it can be a single anomalous event. However, if these 30° are repeated or even exceeded in the most recent years of the time series, and therefore the average summer temperature of a long period of summers increases to, for instance, 28° (two more than the previous average), then one could conclude that there is a warming trend.

In summary, the abrupt variation of temperature in one year may be alarming, but it is not significant. What really matters is the average variation of a series of years, as this would mark a change in the previous trend. This is done with the most representative weather variables (temperature, precipitation, humidity, wind...), based on historical records and current measurements. A good example of what we are talking about would be the average temperature series of the planet calculated by some of the most prestigious climatological centers in the world, such as the Hadley Center in the United Kingdom, the Copernicus Climate Service of Europe or the NASA GISS program. The thermal deviations from the 1901–2000 average allow us to see the trend of the annual temperature variation, although the five-year averages provide a clearer picture of the general trends. The series shown in Figure 2.3 includes data since 1880, showing a clear warming trend from the 1960s onwards, with a particularly clear increase in the period 1980 to the present. 2016, 2020 and 2023 have been the warmest on record, The thermal anomaly, according to the NOAA climate records, reached 1.18° C in 2023, and reached 1.5° C when compared to pre-industrial times.

When the issue of climate change began to be discussed in scientific circles (it would later spread to public opinion), skeptics argued that the observed increase in temperatures was not significant, or, in other words, that the increases could be due to chance or measurement errors. Other possible error factors were also pointed out, such as the presence of the measuring stations in urban areas, which would be affected by the heat island effect (temperatures in cities are higher than in the surrounding area because of the thermal impact of building materials). Proponents of warming argued that these factors were already corrected and that the increase was also observed in ocean temperatures.

Today, few doubt that warming is occurring, as other evidences pointing in the same direction has been added to the thermal increase trends: loss of ice

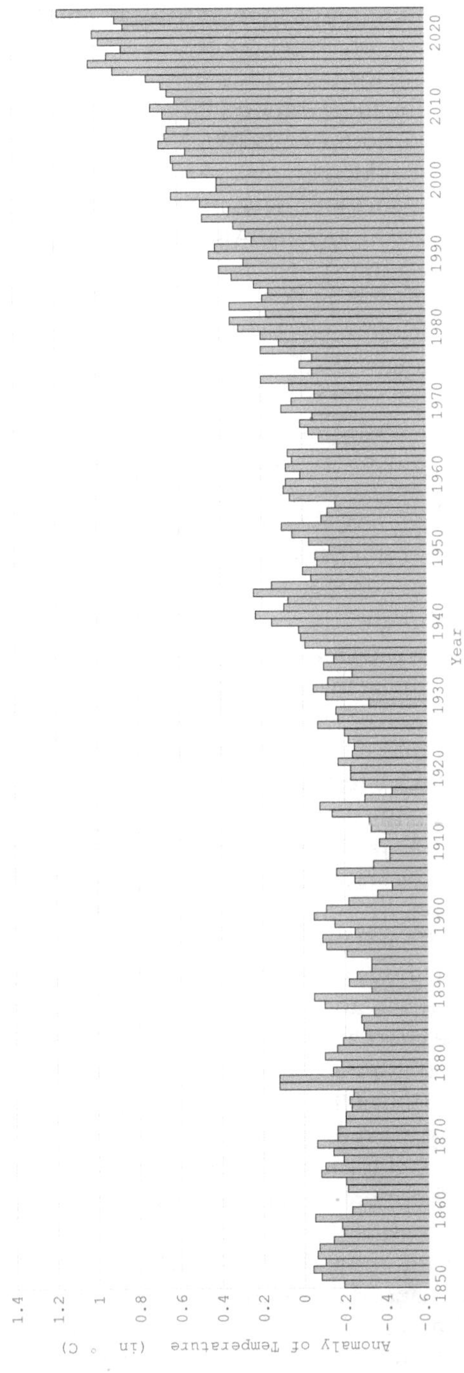

FIGURE 2.3
Global Land and Ocean Average Temperature Anomalies (1851–2021). Base Period for calculation of anomalies is 1901–2000. Data estimated from the NOAA Center for Environmental Information (https://ncei.noaa.gov/access/monitoring/monthly-report/global/202313).

in Greenland and Antarctica, glacier retreats, melting of the Arctic Ice shelf, displacement of species to higher or more northerly areas, rising sea levels, among others. Since so many convergences are observed along the same lines, now critics focus on debating the main cause of the observed changes: natural agents or human actions. The most important natural factors are variations in the Earth's orbit (in three periods of 100,000, 40,000 and 20,000 years known as the Milankovitch cycles), which involve a significant alteration of the solar radiation incident on the planet and have in fact have been the protagonists of past glacial and interglacial epochs. Also noteworthy are the variations in the composition of the atmosphere (emissions from volcanoes, marine or continental aerosols, water vapor), in sunspots, or in the position of the continents. These factors have undoubtedly played a key role in explaining the numerous climatic changes that have occurred in the geological history of the planet. However, we do not now observe today any natural factor that could explain the global increase in temperatures, since the solar radiation reaching the Earth, the atmospheric water vapor or volcanic emissions are, on average, quite stable, not showing any significant trend.

If natural factors are today not relevant, it seems reasonable to think that human activities are behind the thermal effect we observe. The mechanisms are well known, as land cover change and emissions from burning fossil fuels have a direct impact on climate. The conversion of forest to crops or pasture involves changes in the terrestrial albedo (the energy that is reflected by the Earth to outer space), which means that the amount of solar energy absorbed by the Earth changes. On the other hand, the conversion of forests to crops changes the amount of stored terrestrial carbon. This carbon is released into the atmosphere when the vegetation burns or decomposes, mixing with oxygen and thus increasing the density of carbon dioxide (CO_2) in the atmosphere. As for the combustion of fossil fuels, it is also well known that the vast majority of energy used for transportation, heating or cooling, and industrial production comes from burning gas, oil or coal, and all burning process implies the emission of different particles and gases (CO, CO_2, CH_4 and N_2O), among which greenhouse gases (GHG) are a great proportion.

The greenhouse effect in the atmosphere implies that incoming solar radiation passes through these gases and heats the Earth, with more or less strength depending on the angle of Sun rays (which varies throughout the time of the day the latitude and the season of the year) and the type of surface (some covers reflect a lot of solar energy, such as the snow, while others absorb a high proportion: water, vegetation and dark soils). The more energy the Earth absorbs, the more is heated and the more it radiates later. However, this radiation is different from the solar one. It is produced at a different spectral wavelength, which is called thermal band of the electromagnetic spectrum. Part of this outgoing thermal radiation is absorbed by the GHG and goes back to the Earth. The most important and abundant GHG are water vapor, CO_2, CH_4 and N_2O, particularly the former, which is very dynamic (only last a few days), while the others may remain in the atmosphere for several decades. It is well known that the greenhouse effect is very beneficial for the Earth's climate, as otherwise the average temperature of the planet would be around 15° lower. However, the high increase in the atmospheric density of GHG (in the case of CO_2 from about 300 to 419 parts per million in just one century), has been found to be the main driver behind observed trends of climate change. The high increase of CO_2 concentrations is caused by land use changes (particularly deforestation) and mostly by burning fossil fuels, which began in the 18th century with the industrial revolution and is now spreading and intensifying in all regions of the world, as a result of the high dependence of current economic production from fossil energy. Humanity is burning in around 300 years what nature has taken 300 million years to produce.

It is not the purpose of this book to determine the effects and possible solutions to climate change, but it is worth at least pointing out that it is an issue with multiple implications that affect many different aspects of both societies and ecosystems. Rising temperatures affect the balance between liquid and solid water and, therefore, sea levels, the size of glaciers and the water resources available for agriculture and human consumption. Higher temperatures have also an impact on the occurrence of life-threatening climatic anomalies, such as heat waves, extreme fires or catastrophic floods; It also modifies the distribution of species, favoring invasive over endemic kinds, and makes it easier for vectors of certain diseases to jump to other areas, altering the current radii of impact. In addition, the higher concentration of CO_2 in the atmosphere means greater absorption of carbon by the ocean, which consequently increases its acidity (part of the CO_2 absorbed is converted into carbonic acid), which implies greater acidification of the ocean, affecting the marine food chain.

Some of these effects are already being observed as indicated in the Intergovernmental Panel on Climate Change (IPCC)'s reports (2021, 2022); others are predicted on the basis of models that have associated uncertainties and may therefore be greater or less than predicted. In any case, this is a sufficiently relevant issue to have given rise to the largest environmental agreement in history: the UN Framework for Climate Change (UNFCC) signed during the Earth summit of *Rio de Janeiro* in 1992. All states signed the framework and meet annually to analyze climate trends and take measurements to mitigate the problem. Among those annual governmental encounters, the most relevant so far have been the climate summit organized in Kyoto in 1997, where for the first time the most developed countries agreed to establish emission limits by introducing technologies that would reduce the dependence of fossil fuels, and in Paris, in November 2015, where for the first time all countries agreed to establish global emission limits.

The Kyoto Protocol established binding emission reduction targets for industrialized countries. This protocol was not intended to put an end to the problem, but to be a first step in limiting the emissions of the most industrialized countries. Its objectives were very modest and did not involve all emitting countries (only the economically developed ones), in addition to the fact that the main emitter at the time (USA), never ratified it, but at least it served to raise awareness of the seriousness of the problem and to start making efforts in the right direction.

In 2015, the Paris Agreement was signed, which attempted to extend Kyoto to all countries, including the emerging developing ones, but at the cost of greatly limiting the commitments. In fact, it was only agreed that all would send an annual report on national GHG emissions, and that they would include binding commitments in their plans to ensure that global warming of the planet would not exceed 2°C above the pre-industrial temperature (in 2024 the thermal anomaly reached 1.5°C above, being the warmest year on record). However, no mechanisms were established to penalize the non-compliance of this commitment, so that each country's actions can be considered voluntary, with the implementation of these commitments being postponed to successive negotiations. However, first the global epidemic and then the war in Ukraine have hindered faster progress in this process.

In the last years, the efforts to mitigate climate change (by reducing GHG emissions or by capturing those already emitted) have been complemented with policies towards adaptation, introducing measures that will alleviate the negative impacts of the new climate conditions. Ethically, and even economically, mitigation-oriented measures seem more adequate in the long run as they would avoid potentially catastrophic impacts of climate change. Several authors have estimated that investing in mitigation is cheaper than adapting to those extreme climate scenarios (Stern 2007).

Along with mitigation and adaptation measures, other authors, fortunately still in the minority, are in favor of so-called geoengineering solutions, which would involve artificially altering the planet's climate (lowering the temperature through global mechanisms, such as orbital reflectors or stratospheric aerosols). These measures do not modify the root of the problem (an economy based on fossil fuels), but rather its effects (warming), with a very high risk of introducing other potentially much more harmful effects (a sharp drop in the Earth's temperature, which could lead to a new glaciation).

From the point of view of this book, climate change seems to us to be an environmental issue of great interest because it includes a series of aspects that are present, in one way or another, in any other problem affecting the natural environment:

- It shows how environmental problems are multifactorial and that altering one has an impact on the whole: comprehensive solutions must be found.
- It implies a clear responsibility for human action, which requires a response on our part to prevent its occurrence (and not just to reduce its impact).

- It has a global dimension (all countries can potentially be affected).
- Technology is a great ally, but it is not the only answer to the problem. The solutions go deeper: they affect our way of life, where we set our priorities and how we relate to the environment.

After this brief presentation of the interest and importance of environmental problems, we will present the reasons that justify environmental conservation, from different points of view (economic, health, beauty...), but focusing more on philosophical or religious motivations, which tend to generate a greater ethical commitment. Before beginning this review, we will present a brief history of the environmental movement.

Exercise for Reflection

Climate change is undoubtedly the most important global environmental problem nowadays. Being global implies that all nations should take measures to assure an effective mitigation and adaptation to this problem, and being related to the combustion of fossil fuels, the core energy source of world's economy, entails a formidable challenge to find alternatives, particularly at the urgency required. Both aspects of this problem convert climate change into a tremendous complex issue, which involves almost all sectors of economy and requires international agreements of countries with very different degrees of technical and economic development.

For these reasons, climate change is also a very controversial topic, being widely covered by social and general media. No other scientific problem is more debatable by lay people. Despite the growing evidence of the impacts of human-emitted GHG on the observed climate change, recent polls indicate that popular perception about this problem is also growingly skeptical. The paradox indicates that in spite of almost unanimous agreement among scientists (although still specific issues remain under research), the general population considers that climate change is still controversial among scientists. It is not. But, it is indeed among political parties and ideologists that fear regulations derived from a higher reduction in GHG emissions.

The students are asked in this exercise to watch in the social media of their own countries the arguments being used by climate skeptics and try to confront them with those derived from scientific publications. The source of social media could be any of the most popular ones: YouTube, Facebook, X, TED, Instagram… while they can find the most recent scientific studies on similar topics in academic internet engines, such as Google scholar, Academia or Research gate.

The teacher should guide this exercise but dividing the students in groups and asking them first to search about climate change information, then selecting 2–3 media which could be particularly informative. Each group should concentrate on reviewing those media, extracting their main arguments (ideally including skeptics and pro-CC speakers). Each group should focus on 2–3 maximum topics (role of GHG, natural factors, human emissions, trends in temperature, climate-related hazards, etc.) that should be further explored in scientific journals to find out the current state of research on those issues. Finally, each group should present their conclusions to the whole of the classroom for 20–30 minutes, leaving 10 minutes more for a discussion with their classmates.

Note

1 For example, stating that there are one billion people dying of hunger on the planet is more likely to encourage involvement in this serious problem than putting the number at, say, 50 million, but it would not be honest to artificially inflate this number to encourage more action, for the simple reason that we would be misleading public opinion. In this specific example, the most recent estimations by the Food and Agriculture Organization indicate that between 691 and 783 million people suffer from hunger worldwide, with this figure increasing compared to previous years as a result of the impact of COVID-19 and the war in Ukraine (FAO 2023). On the other hand, severe food insecurity causes 148 million children to be stunted. It is difficult to quantify the number of deaths due to lack of food, as this process works in combination with other diseases.

3

Roots of Conservationism

Evolution of Environmental Thought Until the 20th Century

Concern for the natural environment is already present in texts of the early Greek philosophers. For example, Plato in *The Dialogues* (Critias (110b/111d) expresses his concern about environmental degradation: '... *among these mountains, that can no longer feed the bees (...) where not so long ago there were trees*'. This concern about the state of the natural world around us is quite typical of human beings, and shows our dependence on the environment, both for daily survival and to find meaning in the reality that surrounds us.

Historically, the beginning of environmental issues can be traced back to the last third of the 18th century, parallel to the beginning of the industrial revolution, initially located in central and western Europe. This process transformed the structures of the Old Regime towards today's consumer society. The consequences began to become evident in the pioneering countries during the 18th century and, more generally, in the 19th and 20th centuries. Migrations from rural to urban areas quickly extended their size, thus complicating the supply of drinking water and generating a considerable accumulation of waste in the new urban spaces, which led to unhealthy situations. In addition, this expansion was aligned to the construction of industries and infrastructures for transportation, which began to threaten the rural world and the landscape in general.

In contrast with Western Europe, where colonization took place throughout many centuries, this process was particularly rapid in the USA, where transformations from almost pristine areas to croplands or urbanized areas occur massively in only a few decades, covering many parts of the immense territory that extends from New England to California. As a result, environmentalism began in this country in the mid-19th century, as a social movement linked to an innovative philosophical and ethical approach, framing a long dialogue between nature and man, with periods of scarcity and catastrophes. Several specialists have indicated that the individualistic character (self-made, solving one's own problems) that defines American society is largely

DOI: 10.1201/9781003545972-3

linked to this history of colonization, where the pioneer, the one who struggles on the frontier, relies on his own resources to subsist. In this context, nature was frequently seen as an enemy to be fought, the storage from which means of subsistence should be gained. The goal was to 'civilize' (etymologically, to turn into cities) the vast, almost virgin expanses of the country, making them useful for human purposes. This expansion was not only detrimental to the environment, but also to the indigenous peoples who inhabited many regions of the country (it is estimated that there were some 200 nations at the beginning of the 17th century), who lost their lands, their ways of life and in many cases became practically extinct.

The laws of the new country favored family settlements, generally without any environmental regulation, which led to a remarkable degradation, especially in the most fragile lands of the west. The bison, one of the symbols of the country, was on the verge of disappearing, as a result of massive hunting, which indirectly aimed at displacing the indigenous population. The bison crisis was an important impetus for the beginning of the environmental awareness in the country. Along with this colonization in the west, important industrial advances began to take place on the east coast, resulting in an outstanding transformation of the landscape.

The speed and intensity of these changes were decisive for the emergence of a greater conservationist awareness, especially linked to the activity of some thinkers in the first half of the 19th century. A key figure in the generation of the new environmental awareness was Ralph Waldo Emerson (1803–1882), a Protestant minister and founder of the Transcendentalist philosophical movement, which consisted of a group of poets, philosophers, naturalists and writers. The transcendentalists unraveled with their writings and dissemination of ideas the role of each person in society to avoid the depersonalization implied by the industrial era, promoting democracy, freedom and equality as fundamental values, following the ideologists of independence. Although Emerson was not very active in conservationist issues, his essay *'Nature'* had a great influence on other thinkers, as it was the first to propose an equal relationship with nature, with certain pantheistic overtones.

The most emblematic follower of Emerson was Henry David Thoreau (1817–1862), a true forerunner of the environmental movement and a leading figure in the pacifist civil disobedience movements and abolitionism. Thoreau lived for two years in a cabin he built on the shores of Walden Pond in Concord, Massachusetts. He wanted to get away from civilization and live simply, away from the influence of industrialization. His intention was to live in harmony with nature and discover its original beauty, following the inspiration of Emerson's ideas. He captured his experiences and reflections arising from this experience in his book *Walden: Or Life in the Woods* (originally published in 1854). This work, already a classic, has made Thoreau an emblematic figure in American literature. Thoreau poetically advocated a simpler life closer to the environment:

> I went to the woods because I wished to live deliberately, to front only the essential facts of life, and see if I could not learn what it had to teach, and not, when I came to die, discover that I had not lived.
>
> (Thoreau 1854, p. 59)

Along with his enormous influence at the beginning of the conservation movement, Thoreau was also a reference in the defense of the individual against the State, as proclaimed by the transcendentalists. Thus, he voluntarily preferred to go to prison rather than pay taxes to a government that used them to maintain slavery. His book *On the Duty of Civil Disobedience* (1854) foreshadowed the peaceful resistance movements that would later inspire Gandhi and Martin Luther King. The anti-imperialist and anarchist movements also look to Thoreau as a point of reference, although the ideas he expressed, not only in his writings but also in his actions against state oppression, have not always been taken in their proper context.

Thoreau's influence was reflected at the end of the 19th century in the organizations that began to create social concern for the conservation of nature. The first to emerge, which continues today, is the Sierra Club association founded in 1892 by John Muir (1838–1914), another key figure of the environmental movement. Muir came to the United States from Scotland when he was 15 years old and was immediately amazed by the wild and magnificent nature he found on the New Continent. After studying botany, he began a journey of several years visiting different natural areas. He finally settled in California and lived in Yosemite Valley for 9 years (1869 to 1878). He always remained linked to this area, from where he developed works of great interest on glaciology and botany. His best-known activity, however, was the dissemination of environmental awareness, collaborating in local magazines and writing books that contributed significantly to the promotion of conservationism in American society. Muir, who was also greatly influenced by Emerson, was well aware that enjoying the beauty of nature was a vital need for human beings, and therefore more effective action should be taken to ensure its conservation. Muir can be considered the first environmental activist. Like Thoreau, Muir's prose is captivating, and his ideas call for action to protect an environment he considered to be in grave danger. Thanks to his tenacity, he succeeded in lobbying for the creation of Yosemite National Park. The spread of his ideas and his determination would lead to the crystallization of this land protection figure on a national level and the establishment of a network of Federal Parks. The publication of his book *Our National Parks* in 1901 caught the attention of President Theodore Roosevelt, who spent a few days with Muir in Yosemite, asking his help in developing the first environmental conservation policies (Figure 3.1). In fact, during Roosevelt's term, the first U.S. Forest Service was created. He also promoted the development of the country's environmental policy and began to implement the rational management of natural resources.

FIGURE 3.1
Theodore Roosevelt, left, joins John Muir on Overhanging Rock in Yosemite National Park. (https://nps.gov/yose/learn/historyculture/muir.htm)

Less idealistic than Thoreau or Muir, the third great pillar of American conservationism is Gifford Pinchot (1865–1947), who was the first director of the U.S. Forest Service. He coined the term 'conservation ethic' to refer to the need to improve the use of natural resources to better serve human needs. Pinchot showed great leadership in creating a school of forestry that combined the sustainability of the forest with its service to economic growth. In 1895, Frederick Taylor proposed the concept of 'scientific management', which aimed to organize work according to a rational framework that would allow standardization of measures to increase productivity. Pinchot was instrumental in applying these ideas of efficiency to forest management. In times of high demand for timber, such as during World War II, Pinchot's proposals, borrowed from German forestry, to rotate the cutting of forest stands to ensure that timber was always available for industry, became widespread. Since then, practically all forestry schools in the world have been governed by these criteria of efficiency applied to forestry.

It could be said that at this time two somewhat contradictory ideas, embodied by Muir and Pinchot, began to coexist. For the former, it was necessary to promote preservation and conservation for its own sake. For the second, the utilitarian management of resources through rational and technological management was fundamental. Technology as the liberating power of man, which was being forged with progress, also reached the conservationist sphere.

From the same Yale forestry school as Pinchot comes another figure who will have a major impact on environmental ethics: Aldo Leopold (1887–1948). He was a pioneer in ecological restoration techniques and the study of environmental impacts. He proposed to integrate the concept of environmental protection into economic progress. Leopold moved from thinking about management efficiency to advocating environmental conservation as a good in itself. Two years after his death, his personal reflections on the moral and aesthetic dimensions of nature were published in a kind of personal diary entitled 'A Sand County Almanac' (Leopold 1949), which dealt with the seasonal changes he observed on his farm in the Wisconsin countryside. This book is a classic for any American student of the natural sciences and introduces a new way of looking at nature. The third section of this book has been the most cited from Leopold's writings, as it introduced the concept of land ethics, in which he advocated approaching man's relationship with nature in terms of ethical values, not just in economic terms, as was most common at the time: 'a land ethic changes the role of *Homo sapiens* from conqueror of the land-community to plain member and citizen of it' (Leopold 1949, 204).

Just thirteen years after Leopold's book, in 1962, one of the most influential works of the conservation movement was published. It was *Silent Spring*, written by Rachel Carson, a biologist working in the Publications Department of the U.S. Forest Service. This book denounced the indirect effects of DDT (the main insecticide at the time) on bird mortality and human health:

> Every human being is now subjected to contact with dangerous chemicals, from the moment of conception until death. In the less than two decades of their use, the synthetic pesticides have been so thoroughly distributed throughout the animate and inanimate world that they occur virtually everywhere (...) They have entered and lodged in the bodies of fish, birds, reptiles, and domestic and wild animals so universally that scientists carrying on animal experiments find it almost impossible to locate subjects free from such contamination. They have been found in fish in remote mountain lakes in earthworms burrowing in soil, in the eggs of birds-and in man himself. For these chemicals are now stored in the bodies of the vast majority of human beings, regardless of age.
>
> (Carson, 2002:16)

As we noted earlier, society was living in an era of technological optimism in which artificial innovations would provide effective solutions to all human problems. Rachel Carson's work first challenged this belief by denouncing the perverse effects of DDT, which was used as an insecticide for mass spraying of crops and even for household use. Carson's denunciation of this product caused great public alarm and had many consequences. First, the U.S. Environmental Protection Agency was created, one of the first in the world, and the first National Environmental Protection Act, known as NEPA, was passed in 1970. This legislation is the forerunner of environmental law in both North America and Europe, and by extension, the rest of the world. Second, environmental historians understand that this fact was the cause of the introduction of the environmental dimension into the social debate and the generalization of environmental NGOs.

Paradigm Shift

The industrial revolution not only changed the organization of work, but also gave rise to social movements, political parties and trade unions. Industrial progress brought with it a growing perception that the economy was independent of nature, that it was an inexhaustible source of resources, with no limits to growth. We are situated in the so-called HEP paradigm (*Human Exceptionalism Paradigm*), so called because the human being is considered an exceptional creature, exempt from the laws governing other plant and animal species.

In the 1970s, various currents came together to put this paradigm into question. On the one hand, the oil crisis of 1973, which was the first global warning that resources were not infinite; on the other hand, the growing rise of environmental organizations; and finally, the social change that had been brewing since the so-called May of 1968, as a generalized criticism of established values. Thus, a new sociological paradigm was born, known as the *New Environment Paradigm*, according to which the human being is immersed in the cycle of the biosphere as another species, which requires knowing its limits and its responsibilities with the other species, considering human activity in a more holistic, more integrated way.

Thus, throughout the last quarter century of the twentieth century, the main political-economic, social and philosophical ideologies have been reformulated from a NEP paradigm. On the one hand, Marxism has 'dressed' itself green by asserting that the environmental crisis is rooted in capitalism, with its short-term profit strategies involving the degradation of natural resources and human health. On the other hand, liberalism tries to give birth to a 'green capitalism', and turns to economically value the environment and to trust in the regulatory effect of the market to achieve a better quality of life (Martinez de Anguita 2002).

Likewise, the media and educational programs spread interest in nature to society as a whole: informative television programs appeared, and ecology became more accessible and communicative, reaching a large audience. This is the case of some scientist who dedicated their careers to popularizing environmental issues, mostly through TV programs. This is the case of Thomas Lovejoy, producer of the *Nature* program, or David Suzuki, of the *Nature of Things* program, which have been some of the most successful series on television.

Since then, the social impact of naturalist documentaries has played a very crucial role in our environmental awareness. The dissemination of an attractive nature and a better understanding of how natural ecosystems function have been determinant in the generalization of the importance of conserving these places. These programs show that we are part of a cycle of life in which man is just another component, but which is so often the cause of the ecological imbalance we observe.

On the other hand, the growing appreciation of science as almost the sole criterion of objective truth implied a greater interest in ecological research. Faculties of environmental sciences were created, departments of ecology became widespread, and conservation and environmental management gained increasing social acceptance. Among the pioneers of this environmental academic activity is Michael Soulé, who published one of the first articles on conservation biology in 1985 (Soué1985). He proposes a new science based on field studies where ideas used in Leopold's *Land Ethics,* such as integrity or stability, seem to determine the conservation of a specific ecosystem.

Another scientist who has played a major role in shaping our environmental consciousness was Edward O. Wilson (1929–2021), professor at Harvard University and author of numerous popular books. Wilson is considered the father of sociobiology and a pioneer in the field of biodiversity. His ideas are rooted in Social Darwinism, which views human beings as animals slightly more evolved than the chimpanzee. For Wilson, humans do not escape the determinism of the struggle for survival, so in his conception the overpopulation of our species is a threat to the others.

Conservationist Social Movements

The great media impact of Rachel Carson's book marks the moment of consolidation of social presence of groups dedicated to raising public awareness of the environmental limits of human activity, pointing out the negative consequences of uncontrolled growth. The birth of environmentalism as a social and even political movement, with great media and social impact, is situated within this framework, although its roots go back centuries.

Since the founding of the Sierra Club by John Muir in 1882, there has been an explosion of organizations dedicated to the defense of nature, especially in the paradigm-shifting years of the last decades of the 20th century. What began as a movement to study or protect the environment became organized as a social protest against environmental degradation. The sectors calling for an alternative played within the political system, although their goal was not to change policies, but the system itself. Goldsmith, along with the rest of the editorial staff of *The Ecologist* (1972), stated that, 'perhaps the most radical change we propose for the creation of a new social system is decentralization'. This should be based on (i) the extension of democratic participation to all community decisions, which requires that communities be relatively small; (ii) the mixing of urban and rural areas, which facilitates the reduction of transportation, waste, overproduction, or simply the production of unnecessary goods, naturally increasing the quality of products based not so much on the need to survive and expand the market, but on the improvement it will bring to the people of the local community; (iii) the consideration of men and women as individuals, which is facilitated in small communities; and (iv) finally, the reduction of the environmental impact of social organization in small towns and villages. Other authors have also argued that decentralization would be a guarantee of peace because it would eliminate the immense industrial and military targets.

This environmental organizations were increasingly present in the 70s and 80s society, first led by 'Friends of the Earth', a splinter group of the Sierra Club in 1969, and by 'Greenpeace', founded in 1972, helped to form the first environmental parties (Dobson 2016). In Tasmania, Australia, the world's first green parties were founded in 1972, mobilized by opposition to the construction of the Peddler Lake dam. Two years later, in 1974, the *Friends of the Earth* federation decided to stand in national elections in France and Switzerland, winning one member of parliament in Switzerland and 300,000 votes in France, increasing the popularity of the environmental movement in general. At the same time, environmental groups began in Germany to stand for election in 1979 under names such as 'green list', 'multicolor list' or 'alternative list' and achieved electoral successes at regional level. These successes led to the need for a stable national political structure, which was the seed of the 'Green Party', later on extended in other European countries, although only in Germany they have gained a certain political notoriety, particularly after they won 27 seats in the 1983 elections to the Federal Parliament. This has led them to form a coalition government with the Social Democratic Party on several occasions.

Other continents also had a similar political movement, although again much less relevant than in Germany. In Kenya, Wangari Maathai organized the Green Party in 2003, but with little political influence. In the United States, the Green Party gained momentum when attorney Ralph Nader ran for president in 1996 and 2000, although also with little support (less than 3%). In South America, the seringueiros (rubber workers) movement led by Chico Mendez in defense of the Amazon rainforest in the 1980s managed to gain political space in 2003 with the inclusion of a woman, Marina Silva, co-founder with Mendez of the Popular Alliance movement, as Minister of the Environment in Lula da Silva's first government, and who later promoted the Green Party in Brazil in 2009. On this continent, the influence of the indigenist movement in the 2000s has been crucial in the gradual spread of environmentalist ideas throughout the continent. In Colombia, a green party, founded in 2005 under different names, has achieved great success in its short history, winning first the municipality of Bogotá in coalition with other parties, and later on the national presidency in 2022. In 2008, Ecuador took an important step toward what could be called state pantheism by recognizing the Earth, based on the indigenous Inca belief in the deity *Pacha Mama*. President Rafael Correa declared in Article 71 of the Ecuadorian Constitution that 'Nature or Pacha Mama, in which life is reproduced and realized, has the right to full respect for its existence and the maintenance and regeneration of its vital cycles, structure, functions and evolutionary processes'. A few years later, in 2010, Bolivia also passed the Law on the Rights of Mother Earth: 'Mother Earth is the dynamic system formed by the indivisible community of all life systems and living beings, interrelated, interdependent and complementary, sharing a common destiny. Mother Earth is considered sacred in the worldviews of the original indigenous peasant nations and peoples'.

There are currently around 80 green parties around the world with different political relevance, partly related to the actual differences in the electoral systems of each country (Dobson 2016). Germany has a proportional electoral system, which made it easier for the Green Party to be represented, as opposed to the United Kingdom, where the electoral system difficult the entrance of minority parties in parliament. Indeed, this is what happened in New Zealand when it changed its system from majoritarian to proportional thus favouring that the Green Party increased its seats. In general, at present, the proposals of the environmental movement have been taken up mainly by social democratic parties, which has meant that these ideas have advanced more from this political space than from the green parties. With the decline of the left in Europe and the lack of stable space for the Greens, the current strategy is an interaction of ideas between communist and green parties in order not to lose the relevance of the left in Europe. This similar phenomenon is also described by Gudynas (2014) in South America, where he identifies a 'brown' progressivism, that is socially progressive politics that are in favor of undiscriminated transformation of nature, prioritizing economic progress over conservational positions, despite the fact that some Constitutions include a dedicated chapter on the 'Rights of Nature' (Ecuador, Bolivia), unfortunately in blatant contradiction to actual extractives policies. Certainly, in times of economic crisis, conservation goals are often seen as an extra luxury. From this point of view, the economy and social welfare seem to influence the consideration of a green vote.

In recent years, however, there has been a growing skepticism about the real relevance of environmental issues in some political groups, which tend to associate ecological concerns with other social issues, creating suspicions in other citizens that do not share that political agenda. In addition, greater 'noise' comes from social media and certain lobbying groups, which tend to deny the scientific evidence that supports certain environmental regulations, particularly what regards to climate change (Hornsey et al. 2018).

The *Wilderness* Movement

Another conservation movement, and perhaps the oldest in origin, is called *wilderness*. The term *wilderness* refers to the concept of the wild state of nature, which identifies landscapes that have not been altered by humans.

Its historical origin can be found in the German romanticism of the 19th century, specifically in the German Black Forest, although it had a considerable development in the USA, where, as explained above, the direct contact of the settlers with wild nature, as well as the development of the environmentalist thought of Thoreau and Muir, implied an early criticism of the developmentalist mentality. On the one hand, natural landscapes shaped American culture and were identified with the promise of freedom. On the other hand, the advance of the frontier in the process of colonization due to the degradation of resources and landscapes in North America favored the strengthening of the concept of *wilderness*. In 1964 the *American Wilderness Act* was enacted, defining wilderness areas as those areas 'where the Earth and its community of life are untrammeled by man, where man himself is a visitor who does not remain'. In this way, wilderness areas are protected as a natural, spiritual assets, as part of the American cultural heritage.

The exaltation of a pristine, wild nature that excludes compatibility with human presence had for the first time an official entity in the creation of the *Yosemite Valley Preserve* promoted by Muir. From then on, the National Parks movement of the 19th century embraced this *wilderness* philosophy and recognized the cultural, spiritual and scientific value of wilderness areas. For this reason, many of the protected areas in the USA have a zoning where the *wilderness* category leaves certain areas free of any human influence. When attempts have been made to extrapolate this figure to other countries and realities, it has led to many problems, especially in developing countries, where the conflict between the conservation of threatened areas and species and the survival of the local population is great. Guha and Martínez-Alier (1997) have denounced the problems of inequality that imperialist domination causes in this form of conservation.

In Europe, where there was no area untouched by man, the *wilderness* movement took refuge in mountaineering. On the mountain tops is the reduced space of the original beauty of nature. Elisée Reclus (1830–1905), a Belgian geographer, militant anarchist, wrote in 1873 'History of a mountain'. The author, at a delicate moment in his life, after the death of his wife, decided to take refuge in the mountains to find inner peace. The beauty he found in the mountain makes him exalt it as a true sanctuary. This book has inspired the European mountaineering and alpinism movement, which denounces the destruction of these spaces and progress, while exalting the solitude of the mountain as a school of life.

The concept of 'wilderness', which advocates for the preservation of nature in its natural state, also gives rise to the question of whether it is possible to 'restore' nature. In his article 'Faking Nature', Robert Elliot (2006) considers the 'restoration thesis', which suggests that corporations such as mining companies justify their environmental impacts by claiming that the land will regain its value after restoration. Elliot questions whether that value can truly be recovered, comparing it to the distinction between an original work of art and a forgery. Elliot posits that the specific history and origin of each territory cannot be restored, regardless of the efforts made.

Similarly, Eric Katz (2002) asserts that each territory possesses an intrinsic quality, a 'call of the wild', that is beyond the scope of human influence. Consequently, Katz advocates for the concept of wilderness, which opposes the use of technology to remediate degraded ecosystems. This is because the deployment of technology results in the creation of artifacts, which represent a form of domination of nature. In contrast to Elliot's perspective, Katz does not view restoration as a mere imitation but rather as an artifact crafted for human benefit, often without due consideration for the well-being and respect of nature itself. Consequently, Katz posits that there is a moral obligation to resist this domination and to endeavor to preserve wilderness to the greatest extent possible.

In contrast, Andrew Light (2006) states that an individual who has been cured of a disease is not regarded as an 'artifact'. He proposes that the restoration of nature could be viewed as an endeavor to mend the relationship between humans and their environment. Indeed, he argues that restoration is a means of 'freeing' nature from damage that would otherwise impede its autonomy.

The debate surrounding the concept of wilderness, and its implications cannot be fully understood without first addressing the fundamental question of what nature is and what one considers to be natural. These concepts, discussed in depth in the first chapter, are essential to understanding the various positions within environmental ethics and the different motivations for nature conservation that are presented in the next chapter.

Exercise for Reflection

Most social activism about environmental issues is nowadays focused on NGOs and private foundations that aim to increase social awareness and actions about specific environmental issues.

The student may reflect on the values that these NGOs maintain by analyzing their web pages, reflecting on the founding principles they present and the specific initiatives they carry out. The teacher may divide the group in 3–4 working teams, which would perform a first search on the internet on existing NGOs in their own countries-states. Then, each team will select three institutions and perform a detailed analysis on the following items:

- What are the founding principles of that institution? Which are the specific ethical-religious values that it sustains?
- Are those principles consistent with the activities performed by the institution? Suggest possible alternatives in line with its founding principles.
- What is the practical impact of those activities? Could they have more social or environmental repercussion? Suggest alternatives to the type of the activities or the methods to undertake them.
- Present a critical analysis of the principles, social and environmental impacts of the assigned NGOs. Compare among them. Suggest other ways to do similar campaigns.
- Discuss with other teams about your conclusions. Are aligned with other NGOs analyzed?

4

Reasons for Environmental Conservation

Multifaceted Answers to the Environmental Crisis

A few years ago, an association that promotes environmental conservation in Latin America asked through its website (http://ecosofia.org/) what was the main reason that its supporters proposed to conserve nature. The following options were suggested:

Because it is a source of resources that serve human purposes.

Because, if nature is wrong, so is man.

Because man is part of it, and as the only rational being he must take care of it.

Because life and all its beings (humans, animals, plants, landscapes, etc.) have value in themselves.

Because it is a creation of God.

In our opinion, this is a good summary of the motivations that justify a greater effort to conserve nature. They refer to a variety of factors, emphasizing either our own economic interest (the main reason would be human profit), or our well-being (physical or psychological health, to avoid the negative impacts of environmental degradation), or ethical responsibility toward other beings, or intrinsic value (all beings have a value in themselves, independent of our assessment), or our responsibility to God (caring for creation would be part of the divine mandate for human beings).

Although these motivations are not necessarily exclusive, since they can reinforce each other, one's main motivation to care for the Earth may impact one's concern and commitment, since the dominant reasons for environmental concern affect the specific actions that follow the founding principles. Therefore, reviewing the content and consequences of each motivation may help to understand better their implications for environmental action.

DOI: 10.1201/9781003545972-4

Economic Benefits of Nature

Nature is a source of food, drink and clothing, but it is not unlimited: irrational exploitation of these resources will eventually exhaust them to the detriment of all. Therefore, the first reason for caring for the Earth is to conserve our source of sustenance, both for ourselves and for other human beings present and future. From an ethical point of view, this argument sometimes indicates a strong anthropocentric approach, as it considers nature as a mere source of resources to serve human needs, open to being exploited with no other limitations that the well-being of other human beings. This mentality is often associated with those considering conservationism as an ideological discourse that limits individual freedom and precludes economic development, and therefore is basically confronting human interest. This exacerbated anthropocentrism is dressed up in same occasions with altruistic arguments, justifying environmental degradation as an inevitable consequence of economic development and poverty alleviation. Even the relationship between environmental and social deterioration, evident in developing countries, which so often lose their natural wealth for the benefit of richer countries, is overlooked.

We are not, of course, talking about a more primary use of resources, in a survival environment, which has been the dominant one for most societies, from the Paleolithic, with hunting and gathering, through the Neolithic, where agriculture and animal husbandry began, until the industrial revolution, when humans began to become more attached to the technology that transformed the natural landscape. An extreme approach to this gradual distancing is the consideration that any landscape modified by man is more perfect than a natural one. It is appropriate to recall the words of the then President of the United States, Andrew Jackson, pronounced at the beginning of the 19th century by:

> What good man would prefer a country covered with jungles and a few savages to our extensive republic, full of prosperous cities, towns and farms, together with the improvements that industry brings.
>
> (quoted by Lorentzen 2001, 26–27)

This negative concept of nature is reflected, even unconsciously, in the pejorative use of some terms to refer to natural conditions, such as wild or rugged, as opposed to civilized (urban) or cultured (cultivated, agricultural).

If nature were nothing more than a reservoir of raw materials, environmental conservation would be an obstacle for economic development by limiting the means available to produce new goods and services. In this mentality, environmental degradation will only be avoided when regulations require it, and especially when the penalties are greater than the cost of avoiding them.

When environmental legislation is perceived to be too stringent, attempts will be made to circumvent it by relocating production to other countries with less strict regulations. This phenomenon is quite common nowadays, when companies from environmentally sensitive countries have no qualms about installing highly polluting factories in other countries with laxer regulations. The cases of *Union Carbide* in India, where thousands of people died from chemical air pollution in 1968, or of so many mining operations and large polluting industries in developing countries (Congo, Venezuela, Nigeria, Mexico, Vietnam...) are good proof of this trend.

In a more environmentally friendly approach, the economic consideration of nature is reflected in the promotion of sectors that derive an economic benefit from conservation, as would be the case of ecotourism or renewable energies. The former is becoming a source of great importance for some regions and countries, which maintain their natural wealth as a source of attraction for foreign visitors: Costa Rica is a particularly significant example, being the country in the world with the highest percentage of surface area covered by natural parks and the country that receives the highest percentage of income from tourism (around 18% of the foreign currency that enters the country). This sector is growing worldwide. In 2012, it was estimated that ecotourism accounted for 25% of world tourist destinations.

As far as renewable energy is concerned, it is not a matter of analyzing the economic balance, but also its impact on job creation, its role in technological innovation and its importance in the move towards energy self-sufficiency. After the energy crisis caused by the war in Ukraine, few still doubt the importance of expanding the generation of energy from renewable sources, especially photovoltaic solar energy, which has significantly reduced its cost and improved its efficiency, and wind energy, both onshore and offshore.

Another trend in the economic valuation of nature is oriented towards calculating the value of the goods and services that nature provides to humans. This involves considering and valuing resources and processes which are indispensable for the continuity of life on the planet, but which are not usually traded, such as the contribution of vegetation to improve water or air quality or to reduce soil erosion. Economic goods are generally used up when they are consumed: water or wood are used for specific purposes (furniture manufacture). However, water or trees also provide environmental services for the ecosystem as a whole, which are not depleted in themselves but benefit both humans and other living beings. For example, the capacity of a forest to maintain biodiversity is maintained as long as it is active.

Valuing these ecosystem services is complex. On the one hand, some elements are difficult to value. On the other, there is a lack of a market that recognizes the economic value of many environmental services, despite the fact that society values them (Costanza et al. 1997). The benefit generated to society by the existence of the environment through the generation of services (water, soil, air...) 'escapes' from the market transaction itself and this factor not considered in economics is called an externality. An externality is defined as the effect caused by the activities of an agent on the welfare of other agents. Externalities can be either negative or positive. The former occurs when the actions of an agent reduce the welfare of other agents, without any compensation. For example, if an agent pollutes a river upstream and causes diseases downstream, it produces a negative externality if it does not compensate for the damage suffered by the inhabitants of the banks of the lower basin. An example of a positive externality would be a properly managed or conserved forest in an upper watershed, which retains sediments and regulates a pure watercourse that is used by the citizens of the middle or lower watershed of the river. In this case, a natural resource, whether a good or a service, is not compensated by society and can be considered as a positive environmental externality.

This positive environmental externality, which are the benefits derived from having a 'healthy ecosystem', are economic intangibles and constitute the so-called environmental or ecosystem services (Naidoo et al. 2008), or – with a more recent terminology – the natural capital of a particular area (Victor 2020). These amenities can be defined as public services to the extent that all benefit from them, even if some do not consume them individually. For example, carbon sequestration or water storage is not consumed individually, but these services are critical for our health and safety. Within this appreciation of ecosystem services, a framework has been developed to reward those maintaining the proper functioning of natural services. Payments for those services imply a voluntary expense in which the citizen pays for the provision of a particular ecosystem service, under the assumption that this service will be maintained (Wunder 2005). In recent years, this concept has extended considerably, for example, within the framework of programs to reduce deforestation and forest degradation as part of climate change mitigation (REDD programs), which have an important effect on tropical deforestation and biodiversity (Stickler et al. 2009).

In parallel to the strategy of conservation through market mechanisms, there is an innovative legal framework for the use of rural properties with a conservationist approach. These are called land stewardship agreements, defined as a set of legal strategies or techniques through which landowners and land users are involved in the conservation and use of natural, cultural and landscape values and resources. The term was coined in Massachusetts in 1881 by naturalist Charles Eliot, to protect a series of federal lands of great scenic beauty. This first *land trust* spread to other countries in the 20th century. Perhaps the most relevant case is the *National Trust* in England, founded in 1884 to preserve the cultural, natural and scenic heritage of lands outside administrative protection. These *trusts* involve the will to conserve the natural, cultural and landscape heritage involving owners, land users and other public and private stakeholders. A stewardship entity can be formed between very diverse organizations such as legal entities, conservation organizations or citizens associations together with public administrations such as municipalities or consortia that assume the role of the stewardship entity. The agreements generally imply that the owners take the commitment to apply certain protection and management conditions on their properties. In exchange, they receive recognition from the stewardship entity and society, advice on proper land management, annual monitoring of the condition of the farm and the maintenance of the agreement, or the support of volunteers for certain actions.

Another recent tool for nature conservation is the development of habitat banks. This idea is based on the process of compensation and mitigation of damage to natural resources caused by specific projects. The objective of the Nature Conservation Banks is to ensure that there is no net loss of biodiversity, through the preservation of natural resources or ecosystem services. And for its development it is necessary to establish the habitats, species and compensation actions, as well as a system of control and monitoring of these compensatory measures in the project.

Finally, carbon markets have been proposed as an economic strategy to mitigate climate change. These are financial instruments for offsetting emissions by companies or individuals, or for exchanging emission rights between countries, through investments in forestry plantations or in technologies that reduce emissions in the country receiving the investment (for example, renewable generation plants or low-emission transportation systems).

Environmental Health

Another reason for conservationism is to avoid the negative health consequences of the deterioration of nature. Generally speaking, those landscapes, products or processes that have undergone less human transformation tend to be healthier. This is usually the main argument for buying the 'so-called' ecological food, which minimizes the use of artificial chemical compounds in their production or preservation. Generally, the absence of these compounds makes the feeding more expensive, since the cost of agriculture without fertilizers, pesticides or preservatives will likely reduce yields, but people are willing to make up the price difference with standard food by assuming that the latter is less healthy.

It is obvious that all we eat, drink, breathe or touch affects our health in different ways. A healthy natural environment is also a guarantee of human well-being. There are many historical and current examples, where the impact of the abuse of nature on human beings is evident. As mentioned above, the first wake-up call for environmental awareness came because of the publication of Carlson's *Silent Spring*, which denounced the impact of the widespread use of pesticides and herbicides (particularly DDT) on human health. There are many examples of this relationship between environmental pollution and health, from diseases linked to air or water pollution (surface and groundwater), to the impact of denatured foods ('mad cow' crisis, fictitious rapeseed oil, among others), to those associated with mining activities, the proximity to certain industries or the uncontrolled emissions or spills, whether accidental or even deliberate, as in the case of those related to oil transportation. Among the most serious cases are the accidents that occurred in 1984 in Bhopal, India, where an estimated that 15,000 people died from the direct or indirect effects of a chemical leak produced by the *Union Carbide* company, or the nuclear catastrophe of 1986 in Chernobyl, Ukraine, to which 4,000 direct or indirect deaths were attributed (some raise the estimation to 16,000, but there are still no conclusive reports), in addition to more than 300,000 people evacuated. The list could be extended to include with less cases media impact, but of great effect on human health, such as the explosions of fuel depots or refineries, or the daily impacts of river pollution from uncontrolled mining or urban pollution, which in some cities is almost permanent, causing severe respiratory diseases.

One of the environmental social grassroot movements that had higher mediatic repercussion in the USA was the 'Love canal' affair, which converted a simple housewife, Lois Gibbs, in a public celebrity in the country, as she led the resistance to the New York public authorities that want to minimize the health impacts caused by a former heavy-chemical dumpsite on the residents of the area that was built after its sealing (Tal 2006). Twenty years after the closure, a residential area was built around the former dumping site and a school was constructed on top of it. Gibbs move there in 1978 and a few months later she observed health problems in her child. After communicating with other neighbors, she realized that similar problems were found in many other families, which soon were connected with the chemical leakages that were observed in the area. This evidence brought the residents to create a civilian association that pressed the New York state administration to relocate and compensate them for the damage. The movement became an international example of civilian mobilization against environmental health impacts of polluting industries.

Several authors have observed connections between nature quality and psychological and emotional well-being, showing the positive impacts of natural environments on mental health, as they provide inner peace that is difficult to find in contaminated environments. This connection between human psychology and the environment was explained in the 1980s by a theory that Kellert and Wilson (1995) called Biophilia. According to this hypothesis, human beings need contact with nature and other living organisms for their inner balance, because of their genetic connection with them, and is, in short, as another manifestation of evolution. For these authors, advocates of sociobiology or behavioral ecology, the interaction of human beings with the environment facilitates a series of evolutionary advantages. On the contrary, we would be degraded if dissociated from natural landscapes. Biophilia has had a great impact on other fields, such as architecture, coining expressions like 'green design' or 'invigorating environmental design' (Kellert 2005). In addition to improving the energy efficiency of buildings, appropriate orientation, facades with vegetation or organic materials are sought. The idea is that buildings designed according to these criteria will help occupants to feel better psychologically and spiritually.

Educational Aspects of Nature

The educational dimension also provides compelling reasons to preserve nature, since observation and contact with nature helps us to acquire values that make us better human beings. Contemplating nature allows us to understand many processes in our physical environment, which give meaning to both material and spiritual aspects of our existence. The great classical thinkers (Socrates, Plato, Aristotle, Pythagoras, Augustine, Thomas Aquinas, and many others) started from the observation of their environment to forge a logic of reasoning and lay the foundations of scientific knowledge.

This attitude of the classical philosophers is now enriched by sophisticated means of observation, providing a precise vision of the composition and functioning of natural systems. Observing nature allows us to understand better the environment in which we move and how it affects us. Even ordinary people working in natural areas, such as farmers or rangers, can predict changes in weather by repeat observation of certain events, and therefore are able, for instance, to predict when a summer storm or a severe frost may happen. In industry, there has been a recent trend to try to mimic natural processes to create more sophisticated products. An example of this would be the research of some laboratories to imitate the resistance of spider webs in fabrics, or the adherence of geckos to design gloves that allows climbing vertical walls.

As various environmental psychology textbooks point out (Clayton and Myers 2015) people base their behavior on a variety of internal and external factors, and it is the former (those related to knowledge, values, affections and emotions) that usually have the greatest impact on their most important decisions. Among these emotions, one of the most profound is undoubtedly the attraction to beauty, by admiration, wonder. These emotions can be considered immediate, in the sense that they are not linked to reasoning or a searched discovery: they occur without prior warning. They are immediate because they are not mediated by a predisposition. To such an extent that knowledge does not always precede the experience of beauty; nor is it necessary to 'understand' what is admired. Therefore, the ability to enjoy the beauty of the natural world, with the effect that such contemplation has on our psychic and spiritual balance, is a profound reason for environmental conservation. All people who have the means to do so prefer to live in beautiful landscapes: no one chooses to live near a landfill or an iron and steel industry; if they have a choice, they will choose a house with a good view, certainly with wide horizons of natural landscapes. This trend is well exploited by urban development companies, which also value their properties according to the landscape attractiveness of the place where they are located. The higher valuation of areas of special scenic interest would be a simple basis for justifying their protection, avoiding their use for industrial or urban purposes that degrade them.

A sense of wonder at the beauty of nature is often awakened when contemplating a landscape, a sunset, or sunrise in front over the sea.... Romanticism was somehow forged from these experiences, whose aesthetic and spiritual dimensions were the basis of the environmentalist discourse of transcendentalists such as Emerson, Thoreau or Muir. From the latter come these beautiful words:

> Everybody needs beauty as well as bread, places to play in and pray in, where Nature may heal and cheer and give strength to body and soul alike.
>
> (Muir 1912)

American Romanticism also promoted the environmental observation as a learning methodology through the 'nature study movement' with figures such as Wilbur Jackman, John Dewey and Louis Agassiz. They considered that having direct contact with nature was the best way to acquire scientific skills and mental acuity, as well as to develop a certain moral and spiritual discernment (Armitage 2009). Several transcendentalist authors were educated in this kind of pedagogical thinking. Precisely one of these students was Rachel Carson whose fundamental legacy was a didactic of nature, particularly marine biology, through the contemplation of the coast and the tides. In the book *The sense of wonder* (Carson 1998) she speaks of the innate curiosity that every human being possesses. In the context of her experience of spending afternoons with a young nephew on the beaches of Maine, the author pointed out how fundamental it was to cultivate in children this admiration for nature, which enriches their education, in addition to fostering their greater harmony with the environment.

Environmental degradation in the 1970s led to a shift in the conservationist discourse towards a more rationalist, ethically based motivation. Aesthetic issues were largely removed from environmentalism and other motivations, such as environmental health or the balance of ecosystems, took precedence. However, the sense of admiration and joy in the face of natural beauty will undoubtedly be a solid argument for the change in attitudes that we need to promote in order to reverse the current processes of environmental degradation. Too often, attempts are made to encourage this change in citizenship through more or less catastrophic messages, that emphasize the fear of collapse. But most people are paralyzed by fear, especially when the solution is not immediate and does not depend solely on them, when the problems are so great that personal actions seem useless. Fear can paralyze, but we are all stimulated by empathy, by the desire to achieve something better, by love for those who are affected.

An economistic or utilitarian vision of nature, even with the appearance of the good purpose of its conservation, will not solve the environmental and social crisis: it will only patch up and extend the problems because it will maintain the underlying attitudes. It is becoming increasingly clear that environmental conservation must also integrate aesthetic and contemplation, considering nature in a much more comprehensive way, beyond a collection of material processes. As pointed out in a report by the International Union for Conservation of Nature:

> Natural scientists should not be bothered or refuse to consider metaphysical issues into consideration because it would improve both biodiversity conservation and culture, it will not be the scientists alone who will achieve the desired conservation status, but cooperating with poets, artists, philosophers...who influence human behavior.
>
> (Hamilton and McMillan 2004)

Contemplation of nature also helps to elevate our spirit towards meta-material goals. There are innumerable artistic manifestations related to the observation of the environment, in painting, art, literature. It is also associated with the practice of virtue, as St. Basil indicates: 'The contemplation of nature moderates the fever of the spirit and dilutes all insincerity and presumption' (Basil 361–364). Certainly, admiring the beauty and grandeur of certain natural sites arouses the perception of the Sacred, since Creation reflects the beauty and grandeur of the Creator. This explains the abundant presence of many monastic communities — of all spiritual traditions — in places of special natural beauty.

Finally, the importance of nature in children's education is evidenced from the experience of 'Forest Schools', in which very young children are educated in the sense of admiration and curiosity linked to their discovery of natural landscapes. They allow the initiation of the first knowledge in direct contact with the forest environments, enriching their sense of observation and collaboration with other children. Richard Louv has worked on the role of contact with nature in education, describing what he calls 'Nature-deficit disorder' (Louv 2008). He suggests that there is a deficit in children because of the way of life in today's society, in which technology and virtual reality have replaced direct contact with nature. There are several current pathologies in children, such as obesity or apathy, which can be explained by sedentary lifestyles, since children have stopped playing in the street or in the countryside. When children are not encouraged to spend their leisure time in natural settings, they are not able to maintain their attention and have a lower emotional, sensory and intellectual development.

Philosophical and Religious Convictions

Finally, we can also consider as a source of value for environmental conservation the philosophical or religious views that form our web of moral convictions. As we will see in the next two chapters, there are a variety of philosophical and religious approaches to environmental conservation, the result of considering what role we play in relation to other human beings and other living beings.

Simplifying things, and as an introduction to the Chapter 5, we can distinguish two extremes when considering the position of man in nature: anthropocentrism and biocentrism (Figure 4.1). Anthropocentrism emphasizes the centrality of human beings in all their actions, while biocentrism affirms the intrinsic value of all living beings, and places humans on a level of some equality with other species. As extreme versions of both positions, we would have on the one hand strong anthropocentrism, which considers the human individual as the only being with intrinsic value, and consequently any other being would have no more than an instrumental value for us, while radical biocentrism places human beings as just another element of the ecosystem, no more valuable than other species.

Both positions have important ethical limitations. On the one hand, the anthropocentric approach focuses the entire ethical object on human relationships and considers environmental ethics as derived only from human needs or interests. Nature would be manipulable at the will of human beings, independent of its own natural purpose. This vision of anthropocentrism will

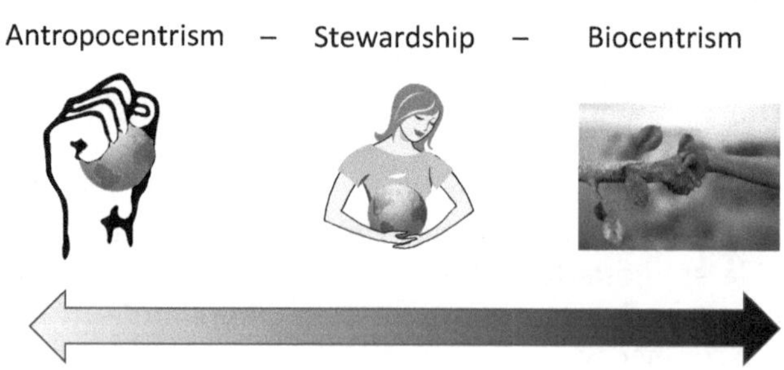

FIGURE 4.1
Range of philosophical positions on environmental conservation.

lead to materialism, with no place for transcendence or total explanation. On the opposite extreme, biocentrism reflects in its ideas the interconnectivity explained to us by ecological science, but by emphasizing interrelationships, it ends up equating everything in the same system. The singular conception of the human being deteriorates, and ends up losing its personal character, with an inability to recognize the human being as 'somebody'.

Exercise for Reflection

This exercise aims to help students to integrate the different arguments that this chapter has presented to promote environmental conservation. First, the teacher may select a specific problem of regional or local interest in which conservation is controversial against other economic or social problems. Examples could be the local impact of a certain industry, which is critical for the regional economy, or the expansion of a conservation area that may trigger social unrest with other local communities (hunters, cattle rangers, real estate developers…).

Divide the classroom into 3–4 teams and assign each group a specific role, within the environmental dilemma. Ask them to document the issue and focus on specific arguments that can be contrasted with other stakeholders with different interest. After a few weeks of study, a debate can be held in which each group presents their arguments for a certain fixed period (20 minutes, for instance), without interruptions. Then, a debate between the groups should be encouraged, with a live discussion on the arguments presented by each group, among them and with the rest of the classroom. The teacher should be the moderator of the debate, assuring a fair tone among the students and the proper respect of the allocated time to each group.

5

Ethical Approaches to the Environment

Ethics and Environment

According to Marcos (2001), 'the idea that environmental ethics is simply applied ethics is false. First of all, because ethics is built from the bottom up, from moral experience to the most abstract principles and not the other way around'. Consequently, the need for environmental ethics starts with concrete facts (environmental problems) and focuses on which contributions the ethical approach would have to solve those problems. Even in a more basic stratum, environmental ethics explores whether there should be limits to our ability to modify the environment, and on which grounds those limits should be established. Since many of our decisions are based on trying to solve dilemmas, that is selecting among alternatives that have both positive and negative effects, ethics helps us to establish criteria to prioritize the different values involved. For instance, considering if nature has an intrinsic value or is simply oriented to serve human beings and whether that service is limited to material goods or also to their longing for beauty and spirituality should impact our choices. As nature entails many different organisms, reflecting on ethical value also implies setting up hierarchies of values for the different components of the ecosystem: one could consider that everything has identical value (biological egalitarianism) or could assume that some have greater value, for example, based on their sensory capacity or their genetic closeness to humans. Ethics discusses how to resolve an ethical dilemma, where conflicting interests and values (rights, impacts, benefits, autonomy...) should be balanced and how environmental risk should be shared among people.

The origin of environmental ethics, as we understand it today, can be traced to the work of Aldo Leopold, who was the first to introduce the term 'Land ethic' (Leopold 1949). He considered that 'the evolution of a land ethic is an intellectual as well as an emotional process.' During the 1960s, the growing concern on the negative impacts of industrial developments on human health and nature, as well as the perception of living on a limited planet, fostered the creation of the first university departments oriented toward environmental

 DOI: 10.1201/9781003545972-5

ethics (Attfield 2018). In 1971 the first Environmental Philosophy Conference was held at the University of Georgia, USA, and a year later, Christopher Stone (1972) published a seminal paper on the ethical basis for granting legal consideration to non-human beings. A few months later, Richard Sylvan opened the debate on the need to create a new ethics with environmental content, and Arne Naess published his essay on deep ecology (Naess 1973), which had notable importance in the environmental debate. Related to environmental ethics, although from a different background, a few years later Peter Singer published his book 'Animal Liberation' (Singer 1975), where he developed an utilitarian approach to the treatment of animal sentience and sensitivity, initiating a proliferation of philosophical works, which today have given rise to the debate on animal rights. At the end of the 1970s the first journal specialized in ecoethics (*Environmental Ethics)* appeared. Since then, research, publication and teaching in these matters have expanded rapidly.

Currently, environmental ethics is considered a branch of applied ethics, associated with the departments of moral philosophy or philosophy of law. It is striking the scarce presence of this subject in the faculties of Environmental Sciences or Biology, which indicates that a separation between scientific and moral assessment persists, as if people and societies as a whole took their environmental decisions based solely on technical criteria. In practice, however, this is not the case: alternatives are chosen according to ethical convictions, habits, economic considerations, social trends or perhaps fear of being penalized. In short, human actions are explained by various factors. Among them, it seems reasonable that ethical convictions should be very important, since they would lead us to solve problems by seeking the optimal moral alternative.

Acting on ethical beliefs involves reflecting on the origin of these convictions and on their effects on the value we place on other beings (humans, animals, plants, etc.). An environmental ethic is built on this reflection and tries to establish solid principles that allow us to make decisions that pursue the greatest good. Naturally, this search leads us to ask ourselves where we can establish these principles and whether they are universally valid or specific to each individual. As Einstein rightly said, relativity applies to physics and not to ethics: if it were so, this world would be very complicated, since everyone could justify any behavior, even the most aberrant, providing that was following his-her own moral criteria.

For ethics in general, and for environmental ethics in particular, it is necessary that the foundations to be as universal as possible, since all human beings, regardless of their convictions, are in relation to the environment and affect it in some way. In this sense, we need to find an ethical theory that fulfills three criteria: 'that recognizes the objective value to living beings, that introduces a certain gradation without discriminating between species, and that does not break the equality among human beings' (Marcos 2001, 111). The challenge is complex to solve, but it is necessary to establish environmental ethics on solid grounds, particularly when considering the great diversity of convictions nowadays. As different authors have indicated, a good common ground would be the renewal of the traditional concept of natural law, so denigrated in the last two centuries, but perfectly suited to the context we are discussing (Turvey 2016).

More than two thousand years ago, Cicero warned us that 'the highest reason is inferred by Nature, which abides those things that ought to be done and forbids those that shouldn't' (Cicero 52 B.C.). In short, the Roman philosopher suggests that living in accordance with nature is a moral criterion and, therefore, a source of moral convictions and legal principles. Here he connects the concept of nature, as a physical reality, with the more philosophical one, identifying it with the essence of beings, which explains why they are what they are and not something else. In this sense, he also affirms:

> ... of all that learned men dispute, nothing, of course, surpasses the clear conviction that we are born for justice and that law is founded on nature and not on arbitrary choice.
>
> (Cicero 52 B.C., book. I, 6)

The same is the approach of several leaders of environmental ethics (Gómez-Heras 2012; Keller 2010), which emphasizes that nature has objective values that generate appreciation, recognition and obligations, as it possesses goods that objectively benefit everyone.

This statement is related to one of the most recurrent themes in environmental ethics, which reflects on the type of value that beings have. Most environmental ethics textbooks divide ethical values into intrinsic and instrumental, depending on whether they consider the value of the being in itself or in its use for other beings.[1] Water has instrumental value for us because it helps us to achieve many objectives: to drink, to refresh, to hydrate what we eat, or even to move around. An animal can also have an instrumental value when considering it as a means for feeding or cultivating the land. The intrinsic value of a being is the appreciation of that being based on their inner properties, regardless of its usefulness to others, which implies that is valued as an end in itself. In this sense, that animal would be valued for its own existence, beyond the interest it may bring to humans or to other animals. In addition to this distinction, Taylor (1986) suggests a third category that he calls 'Inherent' value, meaning the value that we attribute based of its own properties (beauty, historical or cultural importance), rather than its usefulness or commercial value. Taylor differentiates intrinsic and inherent by whether the value is assigned or not by human beings. A great painting that is hidden in a closet has intrinsic value, but not inherent value, as it is unknown by most people. This distinction is not accepted by other environmental philosophers, such as Callicot (1995), who identifies intrinsic and inherent, and Norton (1992), who considers contradictory that value is separated from the valuers, as it would never be achievable a value without conscious valuers.

Whether or not nature and its components have intrinsic value has many consequences and challenges the foundations of a traditional ethics that restricts the moral sphere to human activity (Callicott 1995; Green 1996). If other beings have intrinsic value, it seems reasonable to consider them also as moral subjects, which implies rethinking of our actions towards them, since ethics aims to be normative: that is, ethical principles imply a guide for what to do or not to do. This has very relevant practical consequences, for example, when it comes to establishing priorities in conservation measures or in environmental legislation. If it is argued that the life of an endangered species is more valuable than the life of a human being (who is not endangered as a species), then environmental policies for the conservation of these species could determine the expropriation of human settlements in conflict zones, or even the killing of poachers who attack these species, as it has in fact happened in some African national parks.

To avoid the potential risk of locating the value of nature in the realm of relativistic ethics, with the consequence lack of normative binding, Green (1996) encourages the use of objective assessments, that is the recognition of natural attributes that are valuable to all human beings, such as the role of ecosystems to maintain life, or the dependence of human flourishing from the well-being of the biosphere as a whole.

In the following sections we review how these ethical principles are attempted to be supported in relation to environmental conservation from different philosophical positions (Attfield 2018), which we can place in a continuous range that, as we have seen, fluctuates between strong anthropocentrism and radical biocentrism.

Strong Anthropocentrism

For this approach, only human interests are worth pursuing and therefore nature has a merely instrumental value, serving only to achieve whatever ends people wish to achieve. Therefore, this position identifies the environment with its natural resources, a mere storehouse of raw materials which can be transformed at will, being the only restrictions the technical capacity or the economic viability of the process. Concern for other species would only be justified to the extent that it could satisfy people's needs, whether strictly material, or because of the value they place on those species (for their beauty, diversity, proximity, etc.). In a well-known book on the relationship between species conservation and human needs, William F. Baxter includes a quote that expresses this position quite drastically:

> Damage to penguins, or sugar pines, or geological marvels is, without more, simply irrelevant… Penguins are important because people enjoy seeing then walk about rocks; and furthermore, the well-being of people would be less impaired by halting use of DDT than by giving up penguins. In short, my observations about environmental problems will be people oriented… I have no interest in preserving penguins for their own sake.
>
> (Baxter 1974)

Following his argument, and although Baxter does not conclude so, when people no longer enjoyed seeing penguins, there would be no reason not to eliminate them. We are not suggesting that strong anthropocentrism aims at the disappearance of non-human beings, but that its only argument for preserving them is the benefit they bring us. For several authors, such as Baxter himself, this position is not only not detrimental to the environment, but is the only basis for an environmental ethic, considering the following arguments:

Ethical questions are exclusive to human beings; it makes no sense to apply them to other animals, for whom rights and duties are incomprehensible.

Anthropocentric position is the one most common to human thinking, the one that best fits reality and the only one that is practicable. Humans are free to sacrifice themselves for other species, but they cannot force others to do so.

Humans are largely dependent on the health of nature and, therefore, nature should be protected for the benefit of humans. Environmental prohibitions should benefit as many people as possible.

Considering other species as an end in themselves is very difficult to put into practice. For example, it is difficult to define whether all beings have equal moral importance, and even more difficult to find ways for them to have legal or political representation, or to find ways to objectively establish their preferences.

Environmental problems are generally understood as technical issues, leaving the political decision to take care of the environment at the center. However, in everyday life, when political decisions lead to behavioral changes or restrictions, such as limiting the use of private transportation, protecting certain areas, or increasing the price of gas, social conflict tends to increase. And this conflicts lead to a generalized disaffection with environmental care or even ideological belligerence against it on some persons.

According to various authors, the ideological pillars of the anthropocentric approach to the environment are utilitarianism, technocentrism, economism, and Manichaeism. Ethical utilitarianism leads to the qualification of an action in terms of its consequences. Environmental degradation could be admitted only to the extent that it does not significantly affect other people, since it is a question of achieving a balance between our needs for natural resources and our interest in a non-degraded environment. We might tolerate a certain degree of environmental pollution if it brings with it other benefits (economic activity that reduces unemployment, for example) or avoids other damages (investing economic means that can be devoted to other social needs, for example).

Technocentrism is based on the assumption that the core of environmental problems can be solved through technological development. Environmental problems would be merely technical problems that would be solved by more technological development: if there is a problem of waste accumulation, the solution would be found to build a landfill, to recycle more efficiently, or to use energy more efficiently. Somehow, the starting point is an inherent optimism and a blind faith that man will always find a solution to problems. Human ingenuity will always triumph because, after all, it is assumed that everything is 'under control'. But the truth is that we are very far from understanding the functioning of the planet to properly estimate the secondary effects of certain interventions. Nothing is said about the imbalances that can be caused by technology which attempts to solve one problem, perhaps by introducing another of greater significance. Years ago, it became known about the problems that an exotic herbaceous species of African origin was causing in northern Australia, which was introduced to the continent to stimulate fodder production. This species had higher productivity than the native species, with greater biomass production and height, but the uncontrolled expansion of this species meant that, in the dry season, this grass become senescence and provided more dry biomass than the endemic species to fuel late-season fires. As a result, they were much more voracious, killing tree species that would not have perished with less intense fires. Since this grass did not have an Australian species that would consume it at the necessary rate, the solution afterwards proposed by some was nothing less than to introduce into Australia the African elephants that eat these grasses in their place of origin: a real ecological nonsense.

Finally, anthropocentrism is also related to certain dualistic positions towards nature, such as those derived from Gnosticism and Manicheism, as if it were something completely detached from us, or even hostile to our interests. Nature would be approached from a mechanistic and materialistic point of view, detached from any spiritual significance and therefore worthless. This is the explanation suggested by Derrick (1972), who observes a Manichean dualism in the hostility to the environment of some theorists of the scientific revolution such as Bacon. For Derrick, it is to some extent a pessimistic view of material reality, which considers the world as intrinsically evil, as if human existence implied liberating from material limitations of our body. Consequently, human development would be seen as a battle to impose our dominion over nature.

Ethics of Care and Responsibility

The environmental crisis has revealed the inadequacies of strong anthropocentrism and has made it necessary to modify some of its approaches, resorting to ethical considerations that go beyond human interests. The starting point is a position more open to the intrinsic consideration of non-human beings, based on an ethical responsibility derived from our condition as the only rational beings inhabiting the planet. The environmental ethics proposed by most of the major religions, which consider human beings as stewards of the environment or custodian, responsible before God for the use he makes of the planet's resources, would also be situated in this approach (see especially chapter 6).

Although human needs continue to be considered more important than those of nature, weak anthropocentrisms include various ethical positions that give greater prominence to environmental conservation and grant a certain value to non-human beings. For example, the concept of 'personalist ecologism' (Ballesteros 1995) considers that human beings are part of nature and depend for many processes on the rest of the living beings, although they have a greater excellence. Caring for the Earth would be inseparable from caring for people, since preservation of nature implies maintaining also human existence. In this approach, the main environmental problem is the persistence of subhuman living conditions, and the solutions involve respect for human rights, especially the fundamental right to life and personal development in the family, which would be the place of education for solidarity and care.

The problem of inequality and the possibility of there may be an uncertain future in the disposal of the environment have created a fashion for sustainability. Many of the discussions about sustainable development focus on how to manage resources with an emphasis on social justice or the well-being of future generations. One of the most common definitions of sustainability, that of the famous Brundtland Report to the United Nations entitled *Our Common Future* (Brundtland 1987), describes sustainable development as that which meets the needs of the present generation without compromising the needs of future ones. The basis of this development follows the cost-benefit analysis, which is still anthropocentric, since the objective is to meet human needs, although which are exactly those needs is not evident (the current economy is based on supply many objects that are superfluous), neither which will be the needs of future generations (Valera and Marcos 2014).

Another philosophical approach that approaches human-nature relations from a more conservationist perspective is Hans Jonas' Ethics of Responsibility (1995). This German philosopher adapts Kant's categorical imperative to the relationship between humans and the environment. For Jonas, we are responsible for caring for nature because we are the only rational creatures, capable of knowing the consequences of our actions. We can use the resources of the planet, but we do not have absolute power over it, but rather a commitment to take care of it. For Jonas, all living beings have intrinsic value, not only instrumental value, although he considers that the value of the human being is maximum, because the hierarchy of beings is a function of their possibilities of being: the possibility of value is already a value. Therefore, he proposes a hierarchy that would go from mere existence (being), the capacity for autonomy (living being) to that of rationality (human living being). He emphasizes the importance of also considering our responsibility towards future generations, first with the possibility that they can exist, so that they in turn can be responsible.

Economy of Human Scale

As mentioned above, environmental movements in developing countries have different characteristics from those in industrialized countries, which are more in tune with the vital needs of rural communities. Much of the demands of these peoples have to do with the prevailing economic model, which separates resources from the local environment, in order to include them in a global trade chain that often ends up degrading the immediate resources. One of the first and most lucid criticism of the modern trend towards globalization was provided by the German economist Ernst Friedrich Schumacher, who in 1973 presented an alternative economic model based on the local management of resources, taking advantage of the experiences of the people who live in those places, and which serves the real needs of the people much better than the Western economic system, which prioritizes large investments and infrastructure. He called his proposal 'economy of human scale', within his most influential book entitled 'Small is Beautiful' (Schumacher 1973), with the significant subtitle 'Economics as if people mattered'. In this book Schumacher advocates a technology oriented to serve local needs, decentralized, adapted to the natural and human resources of a given region, which also have a much smaller ecological impact on the environment than large-scale technologies. He proposed to change the concept of mass production by the idea of production for the masses. In this sense, he takes up an expression of Ghandi, who pointed out the problems of sophisticated technologies, that require large investments in infrastructure and energy but have very little impact on the local workforce. In an analysis that, fifty years later with the evident excesses of globalization, has become clearly prophetic, Schumacher pointed out that this economic model ends up widening the financial gap between rich and poor countries, and even within countries, through what he called the 'dual economy', involving immense differences in well-being between persons. To avoid such imbalances, Schumacher suggests promoting a new direction in technological development:

> …, a direction that shall lead it back to the real needs of man, and that also means: to the actual size of man. Man is small, and, therefore, small is beautiful. To go for gigantism is to go for self-destruction.
>
> (Schumacher 1973, 169)

This human-scale economy goes hand in hand with environmental values, as it is better suited to human conditions and is friendly to nature. Hence Schumacher's book has been considered a pioneer in environmental economics. Like other thinkers, he considers that the underlying problem is not so much technical as moral, since a change in attitudes and lifestyles is needed to change the dominant values. This also entails recovering our faith in intrinsic values that go beyond the merely economic:

> If we permit them, this is not due to poverty, as if we could not afford to stop them; it is due to fact that, as a society, we have no firm basis of belief in any meta-economic values, and when there is no such belief the economic calculus takes over. This is quite inevitable. How could it be otherwise? Nature, it has been said, abhors a vacuum, and when the available 'spiritual space' is not filled by some higher motivation, then it will necessarily be filled with something lower, by the small, mean, calculating attitude to life which is rationalised in the economic calculus.
>
> (Schumacher 1973, 123)

Schumacher's book was selected as one of the 100 most influential books of the 20th century. The model he criticized finally prevailed, with an economy oriented towards globalization, but its detriments are still very much in force, after the global impact of the 2008 financial crisis, the CoVID crisis in 2020 or the war in Ukraine and the Near East. In our view, it is one of the few truly alternative proposals to the current economic model, where people and local economies play an almost irrelevant role. The transformation of the current economic model, finding another one that makes it possible to make the integral development of people compatible with the care of the Earth is an urgent task: we cannot continue to feed the current system based on waste and materialistic selfishness, not only because it is clearly unsustainable, but also because of the social impacts it entails.

Cultural Ecology

This approach proposes a change in humanity's relationship with nature, based on a new cultural paradigm. The most prominent representative of this approach is Thomas Berry, a cultural historian, most notably through his book *The Dream of the Earth* (Berry 1988). In his opinion, the obsession to exploit the planet's resources is putting the Earth at the limit of its capacity to sustain us, so that a conceptual change is essential, towards an ecological age, characterized by three principles: differentiation (respect for geological and biological diversity), subjectivity (enrichment of personal experiences, the level of complexity of organisms) and communion (of each reality of the universe with the rest of its components). In Berry's vision, the root of the environmental crisis lies in the loss of the sacred character of nature, produced mainly due to the expansion of science and technology parallel to the industrial revolution.

The author proposes an agenda for ecological change, that would include technology to repair the damage caused by human activity, a new business mentality that considers the long-term benefits of ecological assets, a strengthening of social movements that promote conservation values, and a new educational mentality that would better integrate the intrinsic wisdom of natural processes. In this sense, he suggests a new organization of university education, in successive cycles, that will analyze the diverse biophysical, geological, atmospheric and finally cultural processes that define the life of humanity on the planet: from cosmological and biological evolution through various cultural phases. He considers great cycles Paleolithic, Neolithic, period of the great religions, scientific and technological phase, culminating with an emerging ecological phase, in which we are currently.

Throughout his work, Berry places special emphasis on the connection between the new values of environmental conservation and some of the great cultural traditions, including religious ones, which for him play a very important role in his holistic vision of the problem, from the Christian legacy to that which can be gleaned from the spirituality of some native tribes in the USA, whose religiosity he studied profusely. He links the ecological approach of the concept of bioregion, with a revaluation of the female role in society, since the fertility of the Earth has traditionally been associated with motherhood, as well as with a more communitarian conception of social relations.

Ecofeminism

Feminism, understood as a movement whose objective is to put an end to social discrimination against women, has multiple cultural manifestations. Among these, ecofeminism emerged in the 1970s as a position that sought to interpret the environmental crisis in terms of the marginalization of feminine values. It is understood that violence against women and against nature have a common origin: male dominance.

The term was first used in 1974 by Francoise D'Eubonne and has had different developments always in relation to the affirmation of the ecological dimension of the feminine, coexisting with other social demands such as those concerning health, working conditions and environmental justice. In developed countries, it is seen as a more dialectical tension against male domination, taking on various vindictive tints. The feminist and environmental movements share common goals (equal rights, abolition of hierarchies, etc.). The aim is to promote development based on relations of cooperation rather than domination, and they strongly criticize the capitalist economic model. In addition to this social ecofeminism, there is a more cultural one, that identifies women with the fertility of the Earth, which is present in very diverse cultures (Mother Nature, Pachamama, virgin Earth...). The greater identification of women with conservationist attitudes, related to maternity and natural fertility, is also emphasized. In this link between woman and the environment, more spiritual connections are also raised, exploring the relationship between the feminine, nature and the sacred, in some cases linked to the recovery of primitive religions in which divinity was identified with the Earth itself.

In developing countries, ecofeminism focuses on the unequal impact of environmental pollution on men and women. It is women who bear the brunt of the consequences, since they are the ones who carry out activities that are affected by environmental degradation, such as food preparation, which especially in the most degraded regions, involves traveling long distances to collect drinking water and firewood. According to Lorentzen (2001), women and children in rural areas of Bangladesh spend three to five hours a day collecting firewood and, in Himalayan areas, up to seven hours a day. The same can be said of many other developing countries, where the search for drinking water is even more pressing, as it is estimated that 25% of the world's population has no direct access to it.

A good example of this role of women's active conservationism in developing countries is the Chipko movement, developed mainly by poor rural women in India, who applied Ghandian concepts of peaceful resistance to prevent logging on their land. This movement encouraged women to hug trees to prevent their felling. A particularly relevant voice related to this movement has been Vandana Shiva, an Indian scientist and philosopher, who has been a feminist reference in the struggle against modified agriculture (transgenics) and the negative impacts of globalization (Shiva 1991). Another woman very active in conservationism was Wangari Maathai (1940–2011), a Kenyan environmental scientist, who founded the 'Green Belt Movement', an environmental group that has promoted the planting of more than 30 million trees in Kenya to prevent soil erosion and thus improve the quality of life of women who make their living from agriculture (Maathai 2006). Her work has earned her the affectionate nickname of *Tree Woman*. In 2004 she was awarded the Nobel Peace Prize for 'her contribution to sustainable development, democracy and peace'.

Social Ecology

Social ecology starts from the ecological principles of unity and diversity to propose a social organization based on cooperation and plurality. It has its roots in some thinkers of Marxist and anarchist origin, the most emblematic of whom is Murray Bookchin, one of the first trade union and political activists interested in ecological issues. He wrote his first essay in 1962 (*Our Synthetic Environment*), followed by many others on various political and social issues. Through these essays and the journal 'Green Perspectives' he argues that human domination of nature reflects domination within society, a consequence of the prevailing capitalist-statist order. For social ecology, highly hierarchical societies abuse and damage the environment through their mechanisms for reinforcing control, including some methods of agriculture and resource extraction. For Bookchin, the ideal society approximates an ecosystem, with diversity and harmony, where people are free and decisions are decentralized. In short, for social ecologists the ecological problems reflect unjust structures present in society, since the most affected are the most vulnerable people. Therefore, the concentration of power and wealth and the patterns of consumption should be limited, modifying social institutions for an effective service to the people.

One of the branches of social ecology is currently channeling its activity through the so-called 'transient cities movement', representing a change in the structure of the productive system, an abandonment of large-scale production patterns in favor of local economies, with a certain similarity to Schumacher's proposals. In Europe, the network of ecovillages is beginning to consolidate, which involves an approach to production and consumption that is highly integrated with the natural environment.

Related to social ecology, though with a somewhat different origin, is the environmental justice movement, which began as a grassroot tend for civil rights and against the disproportionate exposure of certain minorities to environmental hazards. This would be the case of landfills or polluting industries that are located next to places inhabited by low-income populations. A paradigmatic example is the proximity of nuclear test sites to indigenous reservations in the U.S. (Keller 2010) or the 'Love Canal' movement previously reviewed (see Environmental Health, Chapter 4).

Environmental justice struggles for a fair distribution of the burden of pollution, considering that both local minorities and less developed countries are often victims of exposure to hazards such as nuclear cemeteries, the construction of incinerators or the laying of high-voltage cables near poor neighborhoods or neighborhoods inhabited by such minorities. The U.S. 'Environmental Justice Act' of 1994 was pioneer in avoiding these environmental biases, as it penalizes spatial discrimination of minorities by requiring a mandatory impact assessment to ensure fair treatment of different socio-economic areas.

In the international context, environmental justice joins the movement for human rights and social participation, especially of indigenous peoples. Many of the demonstrations against climate change are based on this demand, since the countries most affected by the impacts of the problem are those that have contributed the least to creating it, since they are not very industrialized economies, but very vulnerable to the impacts of global warming.

It should be noted that the approaches of social ecology, and to some extent the environmental movements themselves, show notable differences between rich and developing countries. In general terms, we can say that:

> For the environmentalists of the North the purest and best nature is the uncultivated region, the land without human beings (...) The environmentalists of the South think that ecology has a direct relationship with the survival of the human being; in such a way they are concerned about the environment, which cannot be separated from the concerns of social life.
>
> (Lorentzen 2001, p. 38)

In fact, for most environmental organizations in poor countries, basic human needs (avoiding water pollution, soil or firewood losses, access to arable land) are the most decisive motive for resistance and social action. Representatives of this 'Southern environmentalism' would be Vandana Shiva and Wangari Maathai, to whom we have already referred, Leonardo Boff, Brazilian theologian and thinker, who has written extensively on environmental issues (Boff 2012), or Pope Francis, who in addition to an encyclical dedicated to environmental issues (see Chapter 6, Christianity), has denounced the social impacts of environmental degradation in several speeches (Pope Francis 2013a, b, 2020).

Although the origins of social environmentalism have a certain connection with Marxism, in practice it is far removed from the official thinking of communist countries, which are frequently examples of highly hierarchical and centralized societies, where concern for the environment has not been historically a high priority, which explains the great environmental cost of their industrialization. There are numerous examples of this situation in the countries of the former Eastern Europe or in those that formed the USSR, where disasters of the magnitude of the Chernobyl nuclear power plant explosion or the near disappearance of the Aral Sea have only become evident in most recent decades. Similar problems are now found in China, where there are numerous reports of environmental degradation with severe air and river pollution experienced in the most populated cities (Liu and Lin 2019).

Holisms

Holism, from the Greek *hólos* (entire, total), refers to the consideration of ecosystems as a whole, beyond each of its parts. Ecosystems are composed of a set of living and non-living beings, related to each other through exchanges of energy and matter, in a dynamic equilibrium that ensures the maintenance and functioning of the whole. Consequently, for some environmental ethics only the whole should be the goal of conservation, above the species and individuals that comprise the ecosystem.

Several authors have suggested that the pioneer of holistic approaches was Aldo Leopold through his *'Land Ethic'*, where he proposed to extend the field of ethics to human relations with the whole *'biotic community'*. In this sense, Leopold's proposal can be described as holistic, although certainly not with the same meaning that has subsequently been given to this term, since Leopold did not consider all elements of the ecosystem – including humans – to be of equal importance. Leopold was not a philosopher, but a forester, but he started a new ecological reasoning that was closely related to moral values. He advocated for a cultural change that would allow us to value the conservation of the environment with a much more responsible perspective, changing the dominant economistic paradigm with one that includes intrinsic values. In fact, Leopold introduced in *Land Ethics* a new criterion of moral valuation when he stated that:

> The 'key-log' which must be moved to release the evolutionary process for an ethic is simply this: quit thinking about decent land use as solely an economic problem. Examine each question in terms of what is ethically and esthetically right, as well as what is economically expedient. A thing is right when it tends to preserve the integrity, stability, and beauty of the biotic community. It is wrong when it tends otherwise.
>
> (Leopold 1949, p. 224–25)

In short, for Leopold, land ethics implies thinking about the impacts of our actions on ecosystems, introducing their maintenance as an ethical value. There have been many discussions on how to interpret these ideas. One of the thinkers who has self-assigned the role of interpreter of Leopold is Callicott, mainly in two of his works: *Companion to A Sand County Almanac* (Calicott 1987) and *Beyond Land Ethic* (Calicott 1999). In both, he emphasized that Leopold had pointed out the biotic set as ethical criterion, leaving the human being as just another species. In his opinion, integrity, stability and beauty, and not the individual, constitute the focus of moral importance.

Under this interpretation of Callicott, some have tried to explain *Land Ethics*, and ultimately holism, as a kind of sociobiology, in which human beings would only behave ethically because of a greater chance of survival: we would act ethically because that would give us greater evolutionary advantages. This is a Darwinian interpretation of Leopold's work, which does not seem to be drawn directly from his thinking, which is much less radical than that of his commentators. In fact, other recent thinkers such as Brian Norton (1984) claim that Leopold is also anthropocentric since he continues to give more relevance to the human community. Beyond the reflections on his work, Leopold's beautiful and endearing book reflects a man passionate about life and the environment, who wants to bring home the point that man's interests coincide 'with those of biotics'. However, it would be unfair to attribute to him a biocentric approach, which was well beyond the thinking of his time. In fact, Leopold considered that human beings 'have and must' manage the Earth, while he never mentioned animal rights or biological egalitarianism.

Closer to a holistic position than Leopold is the so-called 'Gaia Hypothesis', put forward by the English naturalist James Lovelock (1919–2022) initially in the 1960s and more widely disseminated in 1979 (Lovelock 1979). He chose the name of a Greek goddess to refer to the Earth, as he considered that the planet is a complex living organism, composed of the biosphere, atmosphere, oceans and soil, which has the capacity to self-regulate, ensuring the balance between its various parts to ensure life on the planet. For Lovelock, the Earth is an ethical subject because it has a conscience, feels pain and seeks its equilibrium. Humans should be subservient to the Earth, trying to impact as little as possible on its rhythms. At most our role would focus on being a kind of 'planetary doctor' proposing preventive measures to avoid drastic changes in the conditions of the ecosystem. Humans must recover harmony with Gaia to maintain our permanence on Earth, as we are dependent on the rest of living organisms. The Earth system is composed of many processes and all components are important, man no more than others.

Finally, other holistic authors propose different alternatives for the moral consideration of ecosystems, depending on whether value should be attributed primarily to the species, to ecosystems or to the biosphere as a whole. Within these, in turn, different authors propose the need to establish hierarchies of values. For example, Homes Rolston (1932-...), considering the importance of intrinsic and extrinsic values, indicates four ranks: abiotic entities, which would have minimal intrinsic value and high instrumental value; non-sentient flora and fauna, which would have very weak individual intrinsic value and high instrumental value as a collective; sentient fauna, endowed with high individual intrinsic value and little instrumental value as a collective; and human beings, with the highest intrinsic value as individuals, and the lowest as instruments of the biotic community (Rolston 1988). Lawrence E. Johnson (1992) argues that it is the species, rather than individuals, that are morally significant because they are the ones that ultimately affect the functioning of ecosystems. In this context, only endangered species would be strictly protected in order to avoid the loss of the function they perform in the ecosystem.

Biocentrism

Biocentric visions are varied and have different nuances. Briefly, these visions share a rejection of anthropocentrism and advocate giving intrinsic value to non-human beings. These approaches lead to rethinking the meaning and practical scope of morality, going beyond humans to embrace also embraces environment.

In these approaches, non-human beings also have moral consideration for multiple reasons: for their own existence, associated with the role they play in the ecological balance, for being part of biodiversity; for their capacities as agents, with the ability to self-regulate; because they feel preferences, are capable of assuming obligations or obeying them, because they exhibit intentionality (Vilkka 1997), and because of their capacities as patients, since they are sensitive (they experience pain and suffering), have subjective experiences, and are affected by the behaviors of others. For some authors, this supposes in practice that all living beings deserve the same moral status, which Taylor (1986) calls biocentric equity. In short, in this approach, the human being would not have a superior moral status and any preference based exclusively on belonging to the human species would be considered *speciesism*, for these authors as ethically rejectable as racism or sexism.

Biocentrism often underpins environmental movements and groups, although in practice the actions they take and the ends they advocate in their own media have many connections to human interests, especially on issues potentially affecting health (nuclear or transgenic protests, for example).

The most important biocentric currents are holism and deep ecology. Animalism (animal rights) has some points in common with biocentrism, although they have ideological origins and end up in practical actions that are very different from, and sometimes even contradictory, to those of environmentalist groups.

One of the most widespread biocentrisms is the so-called *Deep ecology*. The term was proposed by the Norwegian philosopher Arne Naess (1912–2009), to describe a system of thought where being, thinking and acting in the world should embrace ecological wisdom and harmony. For Naess, the hierarchy of living beings is artificial, for all have equal value, regardless of their usefulness to humans. Therefore, people have no superior rights to those of other animals, except to satisfy their basic needs. Naess distinguished between two types of ecological movements: the superficial, which is concerned with pollution and consumption, but errs in its rationale for conservation, and the *deep ecology*, which advocates a much more far-reaching cultural change (Naess 1973). Deep ecology should inspire life and culture, for all living beings are intimately connected and possess equal moral status. Humans should not be the center of ethical concern but should be on an equal footing with other living beings. The aim of deep ecology is to change the traditional point of view, to base our decisions on a new moral reflection that also brings with it new attitudes: to eliminate relations of domination in relation to other beings by that of coexistence and cooperation, to propose an active defense of the terrestrial landscapes against pollution and the depletion of resources.

The principles of deep ecology can be summarized as follows:

- The well-being and development of human life and other living beings on the planet have value in themselves.
- The richness and diversity of all forms of life participate in the realization of these values and are themselves also values.
- Human beings have no right to reduce this richness and diversity except to satisfy vital needs.
- The current situation of human interference in the non-human world is excessive and is rapidly deteriorating. The development of human life and its cultures is compatible with a decrease in the human population. The development of non-human life also requires this decrease.
- Therefore, policy must change. Policy changes affect the basis of economic, technological and ideological structures. There will be a major change that will make the present situation very different.
- The change in ideology has its origins primarily in appreciating the quality of life rather than adding to a further increase in living standards.
- Those who advocate the above points must be involved in trying to implement the necessary changes.

For the supporters of this movement, such principles demand a drastic change, both individual and social, ethical and political, because it would be the only way to reverse a self-destructive trend. Therefore, deep ecology is a combination of a philosophical approach with ecological activism. Ecology is the general inspiration for all life and culture. The background of this movement is Darwinian, Platonic, with Taoist, Buddhist and also pacifist roots following Thoreau. The egalitarianism of all species creates a value system suitable for culture and law. It follows the line proposed by Thomas Malthus on the importance of population control and states that the quality of life is opposed to the quantity of people. In this approach, some authors have proposed reducing the world population to one hundred million inhabitants. '*All efforts should be directed at reducing population, using peaceful means*' (Devall, quoted by Ballesteros 1995, 25).

Eco-Resistance

Deep ecology has influenced and continues to influence many environmental movements and has also stimulated philosophical reflection. A radical consequence of this approach is the phenomenon called *ecoresistance* (some call it *ecoterrorism*), made up of groups that consider it necessary to carry out actions of interposition or sometimes confrontation to defend the planet from human aggression. In this context, it is worth mentioning Edward Paul Abbey (1927–1989), who wrote a novel entitled *The Monkey Wrench Gang* (1975), in which he presents four 'anti-heroes' who vandalize the construction of the large Glen Canyon dam. This book is a plea for counterculture and an invitation to acts of sabotage against development that threatens natural life and the environment. Abbey's writings have raised awareness while denouncing the greed and hypocrisy of industrialism. Although he considered himself an heir to Thoreau's peaceful resistance and civil disobedience, he was actually advocating active confrontation to achieve his goals.

An example of this ecoresistence approach is 'Earth First', an anarchist movement in defense of nature and against the social order. Its actions against timber companies in California during the eighties were relevant, later oriented towards more radical actions. They start from the premise that there is a war against the Earth, and they are soldiers in favor of nature, so any self-defense action is morally justifiable. In fact, the group's magazine (http://earthfirstjournal.org/) includes a list of people who have been imprisoned for participating in violent acts, both in USA and other countries.

Of course, there are many other eco-resistance groups that use non-violent actions, such as blocking facilities, activities or communication routes to prevent an action that they consider seriously harmful to nature. This is the tactic of the most visible environmental groups in the media, such as *Greenpeace*, although their activities extend to many other issues.

More recently, the youth movement associated with the figure of Greta Thunberg in her protest against the ineffectiveness of international policies to combat climate change, has given rise to various civil movements such as *Fridays for Future* or *Extinction Rebellion*. The former focuses on peaceful protests (mainly school and work strikes, demonstrations, rallies...) to demand effective actions to ensure compliance with the UNFCC Paris agreement. Their website lists the principles underpinning the movement: non-violence, no property damage, no trash, no profit, no hate, minimize carbon footprint and always refer to science (https://fridaysforfuture.org/). As for *Extinction Rebellion*, they also define themselves as a movement of citizen resistance and civil disobedience aimed at raising citizen awareness of the need to act more effectively in the face of environmental degradation (https://rebellion.global/). They demand the elimination of all fossil energy subsidies and climate neutrality by 2025. They have been involved in a wide variety of protest actions, including some very notorious ones such as those affecting works of art in different museums (throwing liquids, gluing themselves to the works...), together with other climate activists.

Sensocentrism

The extension of moral consideration towards non-human beings has gained great strength in recent years, to the point of drastically modifying our relationship with animals, mainly with those that are closest to us, either genetically (primates) or in our daily dealings with them (pets). This change has had an impact on how animals are used for food, scientific experimentation or cultural expressions (i.e., bullfighting, fox or whale hunting), to the point of placing them at the center of social debate.

Although some classical authors advocated for a moral consideration towards animals, the recent debate on the ethical basis of our relationship with them began with the work of the Australian moral philosopher Peter Singer, and mainly with his book *Animal Liberation* (Singer 1975). According to this author, the frontier of moral consideration has been changing throughout history, extending it to all human beings, regardless of race or sex. In his opinion, just as the barriers that prevented other human beings (slaves, blacks, women...) from having full rights were demolished,[2] it is now necessary to demolish the barrier that prevents non-human beings from being considered as moral subjects, since this would be a discrimination as arbitrary as the previous ones, based in this case not on race, sex or social condition, but on belonging to a certain species (human).

In Singer's opinion, the moral consideration of a being should not be based on the species to which it belongs, but on its characteristics, and specifically on its capacity to feel pleasure or pain or, more importantly, to have that subjective experience. Following a hedonistic and utilitarian line, the goal of ethics would be to maximize pleasure and avoid pain. If an organism cannot experience pain or pleasure, because its nervous system is not sufficiently developed, it would not be a moral subject. For example, for Singer, since trees have no self-perception, it is ethical to cut them down and replace them with others. Under the criterion of being or not being self-aware, Singer establishes a gradation of which beings are or are not subjects of ethical interest. For Singer, self-aware organisms are humans, mammals, especially primates and whales, and these cannot be replaceable. However, animals that do not feel, plants and natural formations, and even human embryos or intellectually disabled persons who cannot suffer are for Singer morally irrelevant in themselves, although he admits that they have instrumental value in providing a pleasurable experience to the lives of others. Accordingly, Singer considers involuntary euthanasia for severely disabled patients, abortion or infanticide to be ethically acceptable, since he establishes the capacity for conscious feeling as the ethical boundary. In the field of animal welfare, Singer proposes that we eradicate the consumption of meat, experimentation with animals or their use as a means of entertainment. He advocates the promotion of animal welfare and the recognition of animal rights, beginning with the species most genetically similar to us.

Another author who argues for animal rights is Tom Regan (1938–2017). In his opinion, the concept of animals as mere objects for our use is behind the suffering and death inflicted on them. Consequently, a comprehensive ethical decision should lead to the complete elimination of such use, both in our food and clothing, as well as in experimentation – even for scientific purposes – and naturally in entertainment. (Regan 1983).

A much-debated issue among 'animalists' is the appropriateness of assigning priority moral values to animals. For some, as in the case of Singer, genetic proximity to humans is a criterion of value, and therefore primates and animals with a more developed nervous system (more capable of experiencing suffering) should be more respected. For other authors, these boundaries are artificial, and the development of the nervous system should not be a barrier to considering an animal as having intrinsic value. According to this second group, establishing these divisions would be equivalent to establishing a new artificial ethical boundary, which they call sensocentrism, and which would not be justified on purely biological grounds, since most animal species have the capacity to feel.

There are also several authors who are critical of these positions, especially from a strictly environmental point of view. For example, Sagoff (1984) considers that animalistic approaches lack any ecological logic, since ecosystems need intra-species violence in order to maintain equilibrium. Predation, which is a fundamental part of ecological balance, is present in nature. If human beings were to take as a moral value the defense of animals, for example, against other animals, all natural regulation would be destroyed, based in large part on trophic chains (herbivore animals eat plants, carnivorous eat other animals), which are essential to regulate the populations of different species.

Despite these criticisms, animal welfare advocacy has spread remarkably, having gained enormous acceptance in most countries. The first animal welfare legislation was enacted in 1967 in the United States with the 'Animal Welfare Act'. Since then, many others have followed, both in North America and in the European Union and other countries. The aspiration of animal welfare groups is to abolish the exploitation of animals used as resources: animal farming, beekeeping and hunting should be banned, as should any experimentation on animals, and we should recognize basic rights for them. In this context, the Universal Declaration of Animal Rights, a copy of the Declaration of Human Rights, adopted by the International League for Animal Rights in London (1977) and subsequently approved by the UN and UNESCO, should be highlighted. A step forward 20 years after this declaration is what is known as the 'Great Ape Project', an initiative signed by Singer, Regan, Dawkins and Goodall, among others. This Project attempts to recognize primates as subjects of rights, on the basis of the genetic closeness with humans.

The three objectives to be achieved in the Great Ape Project for primates are: the right to life; the protection of individual freedom and the prohibition of torture.

Assessment of Different Ethical Approaches

We have presented a brief overview of the main proposals that have been developed in the field of environmental ethics in recent decades. The positions are very varied, depending on the starting points of each author or movement.

To summarize, we can point out that there is a clear diversity between those who think that ethical issues only affect human beings, and those who extend the ethical frontier to other living beings, as objects or subjects. In the first case, our relations with animals are considered to have ethical implications insofar as they affect other human beings (e.g., the owners of those animals) or our own moral virtues (compassion, affection). In this sense, the ethical consideration of dealing with non-human beings has quite ancient roots, which can be found in classical authors or in the sacred texts of most of the major religions.

Much more novel is to consider animals as subjects of rights. The position is attractive because it implies an attitude of openness to the good of others and moral generosity, if we can use that expression. However, it certainly has many debatable points, both on the theoretical level — for example, violating ecological relations between animals, often based on predation — and on the practical level, since applying legal principles to those who cannot exercise them would lead to many controversies that would be difficult to solve.

Another line that allows us to classify positions in environmental ethics refers to the relationship between environmental and social problems. We have seen how various ethical approaches interpret environmental conflicts as an extension of conflicts arising from social differences (the basis of social ecology) or between the sexes (the basis of ecofeminism). To a certain extent, it can be seen that attitudes towards the environment are also a reflection of openness to 'others', and that the willingness to dialogue and to overcome conflicts by giving up some of our own well-being is, in the end, a way of achieving a balance that will ultimately benefit everyone. This is also applicable to other environmental ethics that value the moral component of our relationship with the environment, as a personal commitment that also brings with it a new approach to the economy and human relations. In this sense, although coming from different sources, there is a convergence between thinkers coming from socialist-anarchism and Christian humanism, such as the cultural ecology proposals of Berry or the personalist ecologism of Ballesteros or Schumacher. In both cases, they stress the need to do *'something more'* to solve environmental problems than to entrust everything to the power of technology, on the evidence that a technical science without ethical guidance ends up being contrary to the integral good of human beings.

Exercise for Reflection

In 2007 the then President of Ecuador, Rafael Correa made an international plea to request funding from donors to avoid oil-drilling in the Yasuni National Park, a UNESCO world heritage site, one of the most biodiverse regions in the world. The idea was to somehow compensate the absence of incomings from the oil extraction in the area, which would imply around 700 million US$ in national revenues, by requesting from the international community to fund half of this amount to compensate Ecuador for preserving the natural area and avoiding fossil fuel emissions derived from oil-extraction. This money would be used to foster a more sustainable economic development in the country, now highly dependent on oil revenues, while maintain social and environmental benefits for the area, which hosts indigenous people living voluntarily uncontacted.

The plea, officially named Yasuni Ishpingo-Tambococha-Tiputini (in brief Yasuni ITT) was open until 2013, when it was officially canceled because of the insignificant response of the donors, which only reached less than 0.5% of the expected goal. As a result of the failure of this initiative, President Correa authorized the oil drilling in the region, which has been active since then, confirming the negative environmental impacts that were foreseen (deforestation, air and water pollution, solid waste, disturbance of local indigenous population, etc.).[3]

The initiative of President Correa along with several NGOs and multilateral organizations implied an innovative approach to environmental conservation in developing countries, helping those vulnerable economies to promote a more sustainable form or progress, while keeping their natural values. From the traditional 'whoever pollutes, pays' in this case the motto would be 'whoever preserves, receives'. The project aimed to prevent the deforestation of almost 5 million hectares, while providing an active management along with indigenous communities (some of them living in voluntary isolation) of another 5 million and assuring the reforestation of 1 million more owned by small landholders. Alternatives to the oil industry were envisaged in the region, including ecotourism, health and education services. However, some thinkers criticized this initiative, as President Correa was making the environmental protection of the area dependent on the international community, when he had promoted the inclusion of the 'Rights of Nature' in the national constitution: it seems contradictory to make dependable a basic constitutional right from external actors (Gudynas 2014). The President finally authorized the drilling of a small part of the Park in 2016, which generated large controversy in the country. After several years of unrest and citizen initiatives against the drilling, a referendum was held in 2023 to take a national decision about this undertaking. The majority of voters were against oil exploration (59 versus 41%).

The case is an interesting example of environmental dilemma, an ethical problem where all solutions seem to have negative and positive outcomes. Therefore, we can use it as a case study on how different ethical approaches would favor different solutions. The exercise can be organized around a role play debate between the students, each one defending the position of different actors. First, the students should document the specific case study, reading relevant sources about the different alternatives. Then, the teacher may assign the different roles following students' preferences or, even better, to assign them randomly, so anyone is expected to consider all positions. The debate should follow a certain rule of courtesy and fair dialogue between all partners.

Potential roles: a) representative of the minister of economy; b) representative of the oil industry; c) representative of the environmental NGOs; d) representative of the indigenous peoples; e) representative of minister of environment; f) representative of ecotourism industry. During the debate, the students may select different ethical arguments following the approaches review in this chapter (anthropocentrism, social ecology, ecofeminism, land ethics, deep ecology, etc.).

After the debate is held, the teacher may wrap up the main arguments used and the different perspectives that are commonly held in environmental dilemmas, aiming to reinforce in the students the importance of considering different perspectives and providing ethical arguments that can be accepted by different positions.

Notes

1 Vilkka 1997 also distinguishes between close instrumental and open instrumental, the former being related to primary ends (a forest serves to produce wood) and the latter to alternative ends (a forest also holds biodiversity and facilitates contemplation).

2 In Singer's opinion, this moral consideration of all human beings does not apply to human embryos or fetus, as he agrees with voluntary abortion.

3 To document this case, see Kingsbury et al. 2019, Davidsen and Kiff 2013 and https://wiki.ubc.ca/Course:CONS200/2016w2/Wiki_Projects/Yasun%C3%AD-ITT_Project_Ecuador, last access August, 2024.

6

Religion and Environmental Conservation

Why Does Religion Matter in Environmental Ethics?

Having analyzed some of the philosophical movements that encourage the conservation of nature, we will now review how the major religions have approached the relationship between human beings and the environment and what consequences these approaches have for the moral values they advocate.

Before beginning this review, it is worth reflecting on the relevance of dealing with religious values in environmental conservation. Strictly speaking, they seem out of place, since ethics should be based exclusively on philosophical convictions, which would be independent of religious beliefs. However, we must not forget that any religion carries with it moral principles of action and that, therefore, it is the origin of ethical convictions. Moreover, in environmental ethics, there are other solid arguments that justify devoting a prominent role to this issue.

A first factor to consider is the importance of religion for a large part of the world's population. From the Western perspective, it seems that religious issues are not part of the public debate, since they are aspects reserved for the sphere of personal intimacy. It is enough to take a walk through any town in Europe to realize that this is not so, since religion has multiple cultural and social manifestations, in addition to the strictly religious ones. If this is true for countries with the highest levels of religious unbelief, the question is much clearer in the rest of the world. Although it is very difficult to determine exactly what it means to be religious, since religion is not a binary variable but allows for many nuances, the Pew Research Foundation estimated with data from 2015 that religion remains an important issue for 88% of the world's population, with varying degrees of relevance (https://pewresearch.org/religion/). Therefore, if we are interested in changing our relationship with nature, it seems at least highly advisable to analyze what the major religions believe about the environment.

A second argument in favor of addressing these issues is the magnitude of the areas conserved by religious communities. In very different

DOI: 10.1201/9781003545972-6

countries with very diverse ecological conditions, from great deserts to areas of incredible wilderness, from polar to tropical latitudes, areas of great environmental interest have been preserved thanks to the fact that the populations that inhabited them considered them sacred, or where religious communities were present, generally looking after that territory. Examples of these conservation areas can be quoted from all religious traditions. Recognition of the work of religious communities have had in conservation has led the International Union for Conservation of Nature to create an international commission on the spiritual and cultural values of protected areas (Mallarach 2012).

A third element to consider is the role that religions play in shaping cosmological visions that ultimately influence human ethical behavior. All religions provide a narrative about the role of human beings in Creation and about their relationships with other created beings. Depending on what that role is, the implications for practical action will be of one order or another, assuming logically that believers are consistent with their beliefs. Both the cosmological vision and the moral consequences that derive from that vision have concrete environmental implications. For example, if one considers nature to be sacred, their attitude toward it will be completely different from if they think that God has given nature as a gift for indiscriminate exploitation. Examining how the major religions approach environmental conservation will help us to understand better its environmental implications: What is man's role in Creation, what are his relationships with other creatures, and to what extent is he responsible for them before God, are questions that carry with them an explanation of man's relationship to the environment. If there is moral consistency with these religious approaches, adopting one or the other should lead to relevant differences in how people of different faiths deal with the environment.

But religions encourage not only a cosmological vision, but also moral behavior rules. All religious traditions encourage frugality of life, a modest consumption in the pursuit of meta-material values that are the basis of vital happiness. All of them also have ascetic objectives, which entail practices of voluntary sacrifice, sometimes limited to specific periods (Christian Lent or Muslim Ramadan), and sometimes permanent (abstinence from certain foods or beverages). All this influences people's environmental impact since, as we shall see later, a large part of our ecological footprint is related to our daily consumption habits.

> Man in every society retains a kind of awareness in persons that is different from scientific or technological reasoning. Religion helps to make human beings aware that there are limits to their control over the animate and inanimate world and that their arrogance and manipulative power over nature can backfire. Religion instils the recognition that human life cannot be measured by material possessions and that the ends of life go beyond conspicuous consumption.
>
> (Narayan 2003, p. 26)

Finally, the relevance of religions in the education of people should be stressed, since a relevant part of worldwide teaching is carried out by religious institutions, even in countries with a sufficiently developed state structure to offer this service. The ultimate aim of this educational involvement of most of the major religions is to transmit, along with knowledge of academic subjects, vital values that accompany the intellectual and moral formation of individuals. In this context, the role of the major religions in environmental education would undoubtedly be of great relevance, since it is linked to values that they have traditionally preached (sobriety of life, non-materialistic vision, concern for the common good, etc.).

The greater involvement of the major religions in environmental issues will undoubtedly benefit conservationism by incorporating spiritual values that will have a significant impact on a large number of people who value religious beliefs. It will also help the major religions in their task of seeking harmonious coexistence among different peoples and in fostering interreligious dialogue on the care of the planet. The fruits of this attitude were the Assisi meetings in 1986 and 2011, promoted by the *Alliance of Religions and Conservation* (ARC), who was created by Prince Philip (husband of Queen Elizabeth II) and was active between 1995 and 2019. Precisely, in a synthesis book on the environmental vision of the major religions, which was published by ARC in 2003, the then President of the World Bank underlined the role that the world's religions could offer to the conservation of nature:

> The reason for collaborating with the ARC is simple. The 11 faiths that now make up the ARC represent two-thirds of the world's population. They own 7% of the planet's habitable land, have 54% of the world's schools and institutionally own between 6 and 8% of the investment market. They are therefore very important players in development. They are also the oldest institutions in the world and possess wisdom on how to live and how to keep hope alive, so we need to listen to them and respect them.
>
> (Wolfenson 2003, XI)

It is appropriate to conclude this introductory section by noting that a given religion cannot be encapsulated within a single cosmological tradition. We cannot simply classify a religion as more or less anthropocentric or biocentric, since different conceptions may have been developed in the historical development of each religion. On the other hand, we must not lose sight of the fact that all religions are theocentric, that is, they are primarily concerned with reflecting on the relationship between God and humanity and, from this vision, develop a framework for sustaining relationships with other human beings and with the environment, which gives rise to different traditions that have many points in common, as we had occasion to analyze in another work (Chuvieco 2012).This facilitates a fruitful collaboration between them in the field of environmental conservation.

Within this framework, we will review the main traditions of the relationship between human beings and nature offered by the major religions, with a brief commentary on indigenous religions. We will deal in more detail with Christianity, since it is the most widespread religion in the world.

Indigenous Traditions

The growing consideration of environmental values in Western society has also led to a revaluation of the religious traditions practiced by indigenous populations, which often place a sacred value on nature. This trend has given rise to a new paganism, which rejects institutionalized religions in favor of more primitive religious forms, some of which are now extinct. The term *pagan* appeared in classical Rome to designate the cults of people living in the countryside (*pagus* = village), and therefore did not have the cultural refinement available in the cities (*civitas* = civilization). The peoples on the fringes of the Roman Empire practiced religions closely related to the environment, their way of life being linked to the resources of the environment. Most of these pagan religions revered natural elements (the sun, the moon, certain animals or certain natural sites). In Europe, these traditions remained alive until the conversion of the Slavic, Germanic and Nordic peoples to Christianity, which occurred gradually between the 4th and 12th centuries. The spread of Christianity and Islam in Africa, the Americas and Asia, also affected these native traditions, although in many regions these religions coexisted with indigenous traditions (animism in the case of Africa, Buddhism-Hinduism in Asia).

For many centuries it was assumed that overcoming paganism implied a key element of development, by replacing a reverential fear of nature, with the dependence of its vagaries on food security, with a greater knowledge and better adaptation to natural cycles, based on a rational knowledge of the environment. In short, paganism was considered a primitive religious state, and the liberation from the divine consideration of nature was accepted as an indisputable form of progress. The postmodern critique of the environmental impact of the implementation of a Western cultural system has involved a more critical view of this transformation, advocating the revival of some forms of primitive religiosity linked to the environment, while enhancing the value of the various indigenous cultures, especially in Australia, the USA and Latin America, as opposed to the Christianity imported from Europe.

As far as environmental conservation is concerned, indigenous cultures are certainly very dependent on natural resources, on life cycles, which is why they often present traditions of great environmental value. They offer a communitarian vision with the Earth as a whole, which they consider sacred (although they are not necessarily pantheists) and of which they consider themselves as members, not owners. Some authors speak of a natural mysticism (Berry 1988), manifested in a rich symbolism, rooted in their creation myths, initiation ceremonies, healing rituals, their celebrations... In this sense, the role of shamans, wise men-priests, is very representative, manifesting in various tribal cultures a close connection between sacred powers and natural rituals, in contrast to the mechanical vision of nature that prevails in Western society.

Another aspect to consider in the relationship between indigenous peoples and environmental conservation is that of environmental justice, which leads to respect for the ways of life and lands of the groups that have occupied them for millennia. In this sense, we find equally reprehensible the attitude of those who seek to expel these groups from their lands in order to achieve an *integral* conservation of certain regions (as if human beings were not part of nature), as that of those who do so in order to exploit their resources. Certainly, this second attitude is much more widespread today, constituting in some places (particularly in the Tropics) both a true cultural genocide and an enormous environmental tragedy. Let us remember that indigenous peoples live in areas that contain 70% of the planet's biodiversity and account for a high proportion of its cultural diversity (90% of the known languages are indigenous).[1]

Hinduism

Hinduism is the predominant religion in South Asia and one of the oldest known. It is the majority religion in India, Nepal and Mauritius, with a significant population in Bangladesh, Indonesia, Sri Lanka and Pakistan, totaling about one billion people. There are many different traditions and practices within Hinduism, so it is difficult to make a general characterization. Hinduism does not really have a founder, nor does it have a source of reference for the interpretation of its sacred scriptures. There are numerous interpreters (gurus) who adapt these texts to the circumstances of each moment. The most revered texts in the Hindu tradition are the Vedas: Rigveda, Yajurveda, Samaveda and Atharvaveda, the oldest and most authoritative being the former. Also of great importance are the commentaries to these books, known as Brahamanas (a total of 19 texts), and the main Sanskrit mythological stories, the Ramayana and the Mahabharata, written in the centuries around the beginning of our era.

Hinduism encompasses a wide range of religious approaches, schools of thought, and cultural traditions, ranging from very simple cultural practices (performed mainly in the home) to more elaborate ritual forms in temples, natural sites, and places of pilgrimage. The worship of God is performed through icons, which are considered a tangible link to the divinity, as it is usually based on an immanent concept of God. Rituals usually include vocal prayers (mantras), chanting, and ablutions, which are seen as ways to achieve inner purification (water as a symbol of cleansing). Birth, wedding, and death ceremonies have a high religious content. Pilgrimages also take place, although they are not normative. The most important holy cities are Allahabad, Haridwar, Varanasi (Benares) and Vrindavan.

The Hindu concept of God is quite diverse, including monotheistic, polytheistic, pantheistic and even atheistic schools (Jagannathan 1984). The most frequent is the worship of a single God, who is considered the Creator of the Universe, Almighty and Eternal, but accepting the existence of other minor gods. In various traditions, the incarnation of God in human figures (avatar) is assumed, as well as the existence of lesser spirits (called devas).

Hinduism generally believes in the immortality of the soul (*atman*), although its destiny is not decided at the moment of death, but it may require going through successive reincarnations. This cycle is guided by the good or bad actions performed, recorded in the *karma*, which would be like a moral deposit of the actions that each individual has done. If the karma is not sufficiently purified to enjoy eternal life, it must dwell in other living beings through the transmigration of souls through various cycles of life (*samsara*), until achieving a state of perfect spiritual union with God (Brahman). This implies the full realization of moral destiny (termed *moksha or nirvana*). Hindus do not bury the corpses but are burned in funeral pyres and scatter the ashes in a sacred river. The moral norms (*dharma*) indicate God's will for each person and serve as an ethical code to achieve this ideal state, through various paths (*yoga*).

In Hindu cosmology, the creator god Purusha was sacrificed so that things could be made. Each part of his body formed some compound of nature. This division of his body gave rise to the castes, one of the most striking peculiarities of Hinduism (Jagannathan 1984). The hierarchy of these castes begins with the Brahmins, priests, over whom the god Agni, representing fire, watches over; they are the interpreters of the scriptures. Next in importance are the Shatriyas, warriors and nobles, protected by the god Indra, then the Vaishyas, farmers, merchants and craftsmen, below the Sudra, slaves and serfs, as well as inhabitants of the conquered territories, and finally the Dalit or untouchables. Women adopt the caste of their husbands or fathers. The higher castes can renounce forming a family, or abandon it when it is already settled, to choose a more demanding spiritual path. These pious Hindus are called '*sadhu*', they eat the food they receive and live in a very precarious way, mainly in hermitages or sacred places or go on pilgrimage to different cities, dedicating themselves exclusively to asceticism and meditation.

In Western perception, Hinduism is considered one of the most environmentally friendly religions. Returning to the first version of the film Avatar, which was built around an environmental conflict, it is worth noting that Hinduism underlies several aspects of the film. The title itself takes up a term commonly used in Hinduism to refer to the different ways in which a god can be incarnated, although it is now commonly used in computer applications as well. The sacred value of certain places (especially trees), the reference to the Earth as a mother goddess, reincarnation, and the global interconnectedness of living beings also have Hindu connotations. Even the blue color of the protagonists (the Na'vi) can be attributed to a Hindu evocation, as this is the color traditionally used to represent the gods Krishna and Rama.

Beyond the simplifications with which the Western vision can interpret Hindu culture, this worldview certainly implies a much closer relationship between human beings and nature than we know in the West. This greater 'environmental connection' is partly related to the immanentist or pantheistic conception of some Hindu traditions. The more general ones consider God to be both transcendent and immanent; he is in all creatures, but not identified with them (panentheism). These approaches also vary historically, depending on whether they are primarily nomadic-pastoral, agricultural or predominantly urban cultures (Gosling 2001).

Hindu sacred texts recognize the presence of God in various natural elements (air, fire, water, soil and space) and perceive the divine existence in all living and inanimate beings. In fact, in some texts, nature is considered a physical manifestation of God himself. It is expressed very plastically in this excerpt from the *Srimad Bhagwat*:

> In the kingdom, the rivers are the veins of the Cosmic Person, and the trees the hairs of his body. The air is his breathing, the hills and mountains are his piled bones and the passing ages his movements.
>
> (quoted by Narayan 2003, p. 33)

For Hinduism the whole universe is considered as a living organism, a vast ecosystem, in which each part is intimately linked to the others, from which it benefits or to which it benefits. In Vedic literature the sky is considered to be our father and the Earth our mother, personified in the goddess Bhumi or Prithvi. The sun gives fertility to the Earth, bringing energy and life. In one of the four most important texts, the Atharvaveda, probably written between the 10th–12th centuries B.C., we read:

> The earth, which possesses oceans, rivers and other sources of water, and which gives us land to produce food grains and on which human beings depend for their survival – may it grant us all our needs for eating and drinking: water, milk and fruit'.
>
> (cited by Gosling 2001, 22)

The Earth responds like a living organism to human treatment, being gentle and fruitful when cared for, so we have a responsibility to prevent damage, avoid any kind of pollution, and use its resources wisely. By treating it gently, we will prevent the Earth from becoming angry and its anger from manifesting itself in natural disasters, droughts, floods, and epidemics. Since the Earth is also God's servant, it rejoices when people properly praise Him.

Abundance is considered a consequence of people's moral goodness. Humans should repair any damage they do to the Earth, for example, by filling in any holes they dig to extract resources; otherwise, an imbalance would occur that could cause earthquakes, landslides, droughts.... It is important to avoid air and water pollution, because they are considered the elixir of life. The Atharvaveda lists the elements that purify the environment: the mountain, water, air, rain and fire (Deva Diwivedi 2003).

For most Hindu authors, veganism is a moral imperative (eating only vegetables, no meat, eggs or fish, although they do drink milk of animal origin), linked to the concept of ahimsa, avoiding any harm to another living being: *'You should not use the body that God has given you to kill other creatures of God, whether human or animal or of any kind'* (Yajurveda Samhita 12.32, quoted by Himalayan Academy 2007). According to this interpretation, one cannot eat meat and lead a harmonious and peaceful life because it harms, at least indirectly, other creatures. Respect for animals and trees is considered as a manifestation of religious spirituality:

> Those noble spirits who practice meditation and the other ways of yoga, who are careful with all things, who protect all animals, are the ones who really take spiritual practices seriously.
>
> (Atharvaveda Samhita 19.48.5, cited by Himalayan Academy 2007)

Another aspect that favors veganism within the Hindu religion is the belief in the transmigration of souls. Since a spirit that does not have a properly purified karma must be reincarnated in another person or in an animal, anyone can carry the spirit of an ancestor, so they avoid using them for food. Despite these arguments, not all Hindus are vegetarians. It is currently estimated that between 25 and 40% of the Indian population is vegetarian (Agrawal 2017). It is more widespread to avoid beef, since the cow is considered a symbol of motherhood, of care for others, and is also an image of generosity. It has been shown that the vegetarian diet is more environmentally sustainable, requiring fewer inputs to produce the same amount of food (Key et al. 2006), so this religious practice has also an environmental effect, reducing GHG emissions. However, the wide use of rice increases the carbon emissions of Indian food over other countries using wheat or corn, because of the seasonal flooding of rice impacts its methane emissions (Pathak et al. 2010).

FIGURE 6.1
Image of Mandala, Kolam and rice flour (Pixabay, free use).

Hinduism considers that the human being is not independent of his environment. It is not a worship of nature as such, but rather worship of a whole that transcends the human being, although there is certainly evidence of worship in sacred places, especially forests, but also in other rites: Kolams, drawings with natural products (Figure 6.1), baths with plants or marriage with trees.

In some Hindu texts it is assumed that the human being is higher in the hierarchy of beings, while others hold that all forms of life, human and nonhuman are of equal value, and all have the same right to exist (Narayan 2003). Humans are part of nature, which in turn is part of their own identity, so that without it they would be lost. There is a particular bond with trees, which are considered the source of life and human existence, especially by rural communities. Gosling (2001) describes a myth of the Didayis in the state of Orissa (southeastern India), in which God decides to recreate the world, destroyed after a flood. In that new world there were no trees, so people could not cook, build houses or find shelter. When God saw their suffering, he added trees to creation. For other traditions, trees and vegetation are considered a representation of the god Shiva, who drinks poison (CO_2) to rid people of it and gives them back the elixir of life (oxygen) (Deva Diwivedi 2003).

The best-known conservationist movements in India, the Chipko and Appiko movements, can be seen as continuations of Hindu cultural and religious traditions rather than Eastern adaptations of the Western environmental movement. Although both movements began in the 1970s, their roots go back to the peasant protests against the abusive exploitation of the Himalayan forests during British colonization, which suppressed large areas of native forest that were cleared in favor of commercial species, the cultivation of land and the expansion of the railroads in the late 19th and early 20th centuries, which in turn displaced the traditional uses of local communities. The new Indian State that emerged after independence continued these practices in favor of the rapid industrialization of the country, disregarding the interests of the local population. It is in this context that the Chipko movement was born, as a protest against extensive deforestation in one of the northeastern states of India and its impact on the massive floods of 1970. The term Chipko (which literally means '*to stick to*') comes from the fact that farmers used to protect trees with their bodies to prevent them from being cut down. Similarly, the Appiko (literally, 'hug') movement in the southeast of the country in 1983, involved peasant women hugging trees to prevent them from being cut down. Both movements take on the pacifist philosophy of Ghandi and have a strong religious focus. Other high-impact environmental movements in the country have focused on improving water management (Anna Hazare).

The extent to which these religious ideas have been put into practice in land-use management in India would require further study. The conservationist approaches currently present in the country are a legacy not only of these religious traditions, but also of the proposals of the political reformers who brought independence to the country, mainly Ghandi and Vivekananda, as well as the influence of some scientists (particularly the physicist Bose), who combined the Indian cultural heritage with the models coming from the British colonization.

In spite of the theorical friendliness of Hinduism towards the environment, India, being the paradigmatic country for this religious tradition, is far from having solved environmental problems. The fast population growth (from 400 m inhabitants in 1950 to more than 1400 m in 2024) has created many difficulties for economic development, implying the reduction of communal lands and forested areas, with little control of industrial (air and water) pollution and uncontrolled urban sprawl, with a high proportion of urban population living in slums. Recent political trends have favored nationalism, with an increase concern about its impact on religious and racial minorities.

Buddhism

It is difficult to classify Buddhism as a major religion because strictly speaking, Buddhists do not believe in God or in the immortality of the soul. For many, Buddhism is more of an existential philosophy, a practice of inner perfection through meditation, along with ethical attitudes and the exercise of self-control. However, several Buddhist schools do admit the existence of different divinities, and almost all of them have temples and devotional practices.

The geographical and cultural origin of Buddhism has many parallels with Hinduism, since it was born in India and had its main development there until its expansion north of the Himalayas. Although there is historical certainty about the existence of Gautama Buddha, originally called Siddhartha

Gautama, his biography has several divergent interpretations. It is known that he was a spiritual master who lived in northern India around the 5th century B.C. Originally trained in Hinduism, he evolved through various ascetic practices until he found an ideal state in which he was able to transcend suffering (reach Nirvana). This moment is known as *enlightenment* or *awakening* (hence the term *Buddha*), and Buddhism would then be the transmission of the principles that allow every human being to reach a similar state of enlightenment. The first disciples of Buddha spread these ideas throughout northern India, initially, and then throughout the rest of the continent. Buddhism is currently practiced by some 500 million people, mainly in China, Japan, Thailand, Vietnam, Myanmar and other countries in the Far East, although it is also present in some intellectual circles in Western countries (Jagannathan 1984).

There is no such thing as a soul or a permanent 'I', but a rather an evolution that takes place through successive *'rebirths'*, either by transmigration or by reincarnation, until the state of spiritual perfection (*nirvana*) is achieved. To achieve this state, a series of ethical principles are proposed, based on meditation, respect for others and what is called 'the middle way', that is, the practice of an inner asceticism that seeks a balance between very strict practices and other more relaxed ones.

The vital foundations of Buddhism can be summarized in four principles, which are called the four noble truths:

- Suffering exists.
- The cause of suffering is the false belief in our own importance.
- The way to overcome suffering is to eliminate the desires, the cravings that cause us dissatisfaction.
- To achieve this overcoming, it is necessary to carry out the eight practices of the 'noble Path': right view or understanding, right thought or intention, right speech, right action, right livelihood, right effort, right mindfulness and right meditation. Those who follow this path of life become *'arhats'*, saints, who no longer have unfulfilled desires and do not need to be reborn, a new birth.

Buddhism is considered one of the spiritual positions closest to environmental conservation, although in this field it has many points in common with Hinduism. Human life is part of a whole cycle of life, as one more element, so it is assumed that humans have no particular pre-eminence over the rest of living beings. The Earth germinates naturally, but human greed leads to division and property conflicts that degenerate into violence and destruction (Pasad 2003). The moral actions of humans also affect the biological and physical balance of the environment; when human morality declines, famine, epidemics and violence increase in parallel, while the renewal of moral values increases prosperity. In one of the oldest Buddhist texts, this is illustrated by the following example:

> Self-growing rice appeared on earth and, out of laziness, people began to hoard food rather than collect each meal. As a result, the growth rate of food could not keep pace with demand. Therefore, land had to be divided among families. After private ownership of land became the order of the day, the greedier people started robbing from others' lands. When they were detected, they denied that they had stolen...The richness of the earth diminished, and self-growing rice disappeared. People had to till the land and cultivate rice. This rice grain was coated in chaff; it needed cleaning before it could be eaten.
>
> (Digha Nikaya, cited by Thakur 2003, 58–59).

The four principles of Buddhism listed above also have an environmental interpretation. On the one hand, the false belief that we are important leads us to lose our connection with the environment, creating the illusion that we are not part of nature. This disunity can lead us to exacerbate our greed which, together with our technological capacity, ultimately explains the pollution of the seas, the air and the Earth itself. On the other hand, overcoming this greed would allow us to live frugally, reducing our impact on natural resources and making the most of what is available. Buddhism advocates a non-aggressive use of the Earth's resources. Another passage in the Digha Nikaya says that man should be like the bee, which uses the nectar of flowers to produce honey, but does not destroy their fruits.

Buddhist ethics are closely associated with nonviolence. Buddhism is against any animal sacrifice, as well as the killing of any living being. According to some interpretations, Buddha also preached against the destruction of any life, even for self-maintenance. However, there are different Buddhist schools in this interpretation. For example, in Tibetan Buddhism they consider that since all life has the same value, if it is necessary to kill animals for food, it is better to do so with large ones rather than small ones, since they allow more people to be fed.

The practical translation of these approaches is not very clear, since a good part of the Buddhist-majority countries are now immersed in a vast process of environmental transformation, most notably China, with very serious environmental problems. Historically, it is of interest to highlight the government of Emperor Asoka the Great, who ruled over almost the entire Indian subcontinent between 269 and 232 B.C. After his conversion to Buddhism, he extended its practice throughout Southeast Asia promoting the construction of roads, hospitals and the reforestation of vast areas, and is therefore considered the closest ruler to Buddhist principles of territorial management.

With respect to current environmental policies, Buddhist countries do not show better environmental indicators than countries with other religions and similar standards of living (Chuvieco, et al. 2016), The case of China is particularly evident, with serious environmental problems linked to the massive transformations that the country has undertaken in the last 50 years. Even though China is not officially a Buddhist state, its population is culturally rooted in Buddhism, but this does not seem to impact environmental practices, which are not particularly exemplary. Bhutan is usually cited as an example of environmental-oriented Buddhist country, mainly because of the original proposal their governors made of a human development index based on indicators of happiness, although its real impact has been debated (Masaki and Tshering 2021).

Judaism

Judaism is a religion and culture that defines an extraordinarily resilient people who have been able to survive as a small minority in often hostile societies. It is one of the oldest monotheistic religions, the forerunner of Christianity. It has no founder as such, although Abraham (c. 1900–1700 B.C.) is considered the 'father of the faith'. Besides him, the role of Moses, liberator and legislator (c. 1300–1200 B.C.) and David (c. 1000 B.C.), the monarch who brought the kingdom of Judah to its maximum splendor, including an area close to what is today the state of Israel, are particularly relevant. The Jewish sacred scriptures are grouped around the books of the Law (*Torah*), the prophets (*Nebiim*) and the writings (*Qetubim*). They are essentially the same books that make up the Old Testament of the Christian Bible, except for a few that Jews do not consider canonical. In addition to the Holy Books, also very relevant to Judaism are the Talmud, formed by a series of commentaries of prestigious rabbis on the sacred scriptures. Some branches of Judaism also highly value the *Kabbala*, mystical writings of difficult interpretation.

The Jewish religious life revolves around some rites of the annual calendar, in commemoration of its historical traditions, like the *Shabbat* (weekly celebration of the day in which God rested after the Creation), *Pesach* (Passover: remembers the night in which Jews left Egypt under the leadership of Moses), *Shavuot* (delivery of the *Torah* in the Mount Sinai), the *Sukkot* (commemorates the forty years of wandering by the desert) or the *Yom Kipur* (day of the great atonement or inner renovation). A good part of the religious acts take place in homes, although they also gather around synagogues for the reading and commentary of the Scriptures.

Judaism has not considered environmental conservation as a priority issue, and there are few examples in the Jewish tradition, other than the Bible itself, that deal prominently with relations with the natural environment (Tirosh-Samuelson 2002). Undoubtedly, the recovery from the trauma of the Holocaust, the birth and survival of the state of Israel, relations with Islam or Christianity, or pluralism within Judaism itself, can be considered much higher priority issues for Jewish thought in recent decades (Waskow 2000).

In a summary of Jewish traditions on environmental conservation, Schwartz (2002) distinguishes four positions: superiority, human beings are above Creation, which serves his needs; delegated dominion, human beings must give an account to God for his treatment of other creatures; admiration, humans are in awe of Creation and fails to understand it; and contemplation, human beings encounters God through nature. As we shall see, these are also traditions that can be applied to Christianity, since the cultural trajectory of both religions has many points of connection.

The tradition of man's superiority over the rest of Creation is clearly defined in the first chapter of Genesis, one of the most frequent quotations to show the supposed anthropocentric character of the Judeo-Christian tradition:

> God blessed them, saying: 'Be fertile and multiply; fill the earth and subdue it. Have dominion over the fish of the sea, the birds of the air, and all the living things that move on the earth'.
>
> (Genesis 1:28)[2]

As we will see below, this quote can be interpreted in a more environmentally-friendly way, but it has also been used to support absolute dominion. There are other passages of Scripture that show this same tradition. For example, Psalm 8 (4–9) states:

> What are humans that you are mindful of them, mere mortals that you care for them? Yet you have made them little less than a god, crowned them with glory and honor. You have given them rule over the works of your hands, put all things at their feet: All sheep and oxen, even the beasts of the field, The birds of the air, the fish of the sea, and whatever swims the paths of the seas.

However, this tradition of dominion is complemented by others that are much closer to conservationist approaches. Among these is what we might call 'delegated dominion', or environmental stewardship, which considers the task of dominating as a subsidiary mandate, which requires human responsibility to care for the Earth, as a gift for which we must give an account. In this tradition, human dominion becomes a care for other creatures, a reflection of God's provident attention for all Creation. This tradition of environmental stewardship appears in the second chapter of Genesis. As is well known, it corresponds to a complementary account of the Creation of the world, where God after creating man '... placed in the Garden of Eden, to till it and keep it' (Genesis 2–15), offering a vision closer to care than to the absolute dominion that would seem to emerge from the first chapter. In short, the Earth does not belong to man, but he receives it as a mission from God. This is quite evident in the rules included in Leviticus concerning the redemption of land and animals during the Sabbath periods:

> Speak to the Israelites and tell them: When you enter the land that I am giving you, let the land, too, keep a sabbath for the LORD. For six years you may sow your field, and for six years prune your vineyard, gathering in their produce. But during the seventh year the land shall have a

> complete rest, a sabbath for the LORD, when you may neither sow your field nor prune your vineyard. The aftergrowth of your harvest you shall not reap, nor shall you pick the grapes of your untrimmed vines in this year of sabbath rest for the land.
>
> (Leviticus 25:2–5)

The basis for these prescriptions is indicated in the same chapter of Leviticus, when it states that 'The land shall not be sold in perpetuity; for the land is mine, and you are but aliens who have become my tenants. Therefore, in every part of the country that you occupy, you must permit the land to be redeemed' (Lev 25:23–24): in short, the land belongs to God and it is up to man to use it as a responsible steward. On the other hand, there are several passages in Scripture that instruct human beings to be kind to other creatures, such as this text taken from the book of Deuteronomy:

> When you are at war with a city and have to lay siege to it for a long time before you capture it, you shall not destroy its trees by putting an axe to them. You may eat their fruit, but you must not cut down the trees. After all, are the trees of the field men, that they should be included in your siege?
>
> (Deuteronomy 20:19)

In short, man is asked to be careful, even in time of war, with the useless destruction of other living things, particularly if they provide fruit to other beings. Within the framework of the non-biblical Jewish tradition there also appear writings that show this same character of delegated stewardship of creation. For example, in the *Ecclesiastes Rabbah*, we read:

> When God created the first human beings, God gave them the Eden Garden and said: 'look to my Works! See how beautiful they are. I have created them for your own good. Do not destroy my World, because if you do so, there will be nobody to repair it'.
>
> (quoted by Schwartz 2002)

The third Jewish tradition on the relationship between humans and nature that Schwartz points out is that of radical admiration, which calls for s an approach to Creation from awe for its complexity and beauty, based on a deep inner humility. It is present in some books of wisdom, such as Ecclesiastes and the Book of Job. The latter shows man's ignorance of the world around him:

> Where were you when I founded the earth? Tell me, if you have understanding. Who determined its size; do you know? Who stretched out the measuring line for it? Into what were its pedestals sunk, and who laid the cornerstone, while the morning stars sang in chorus and all the sons of God shouted for joy?'.
>
> (Job, 38, 4–7)

Such ignorance should eliminate human pride, changing his attitude to a more reverential one towards the surrounding environment. For medieval Jews, knowledge of the environment was one more means of knowing God and being in closer contact with Him. The study of natural phenomena is not detached from the knowledge of the Law (*Torah*), and it is not considered that the revealed truth can confront the truth known through the observation of the outside world. In this line Maimonides (1134–1204) tried to emphasize the interest in the observation of nature, also as a way for the rational demonstration of the existence of God, following in this sense the Aristotelian tradition that would later be taken up by Christian thinkers such as Thomas Aquinas (1225–1274).

Finally, the last Jewish tradition that Schwartz comments on is natural mysticism, which leads to contemplating the world as an extension of God, emphasizing the contemplative meaning of communion with nature. This tradition is present in Hasidism, which is a mystical current within Judaism. As an example of this tradition, he cites a story attributed to Baal Shem, the founder of Hasidism, which shows how the material and the spiritual are not separated:

> Baal Shem once asked a prominent sage about his relationship to prayer, 'What do you do and where do you direct your thoughts when you pray?' He replied, 'I bow to everything that has an individual vitality, which is present in all created things. For in each and in all created things there must be a vitality that derives from the divine influx. I unite with them when I address my words to God so that prayer penetrates into the highest regions'.
>
> (Schwartz 2002, 101)

In addition to these traditions, another very important element in the Jewish tradition with respect to the environment is fidelity to monotheism. An overly spiritual relationship to nature could be mistaken for a form of idolatry that considers created beings as an extension of the Creator. There are many texts in the Jewish Scriptures that explicitly reject this possibility. For example, Moses delivers the commandments he receives from Yahweh to the people with these words:

> You shall not have other gods besides me. You shall not carve idols for yourselves in the shape of anything in the sky above or on the earth below or in the waters beneath the earth; you shall not bow down before them or worship them.
>
> (Deuteronomy 5:7–9)

We must not forget that the monotheism of the Jewish people was an exception in the cultural environment where it developed, so it was an imminent temptation. The book of Exodus recounts one of the first idolatries of the chosen people, when they built a golden calf in despair at the delay of Moses on Mount Sinai (Exodus 32:1–10). This tendency to seek images of God in the material creation is severely criticized in the Hebrew Scripture as a form of idolatry.

Modern Judaism began to take a greater interest in environmental issues after the creation of the State of Israel, when for the first time its vision of the universe could be translated into legal precepts applicable to a country. Certainly, the initial concern was the very survival of the State itself, in constant conflict with its neighbors, and the conservationist agenda has often been sacrificed in the face of pressure to colonize new territories, even in areas with very fragile environmental conditions. Conservationist approaches are rarely related to religious motivations, and the close connection with the territory is more linked to a nationalist approach rather than to a broader environmental interest (Waskow 2000). In other countries, especially the United States, Judaism has developed a more conservationist vision, although almost always driven by non-religious motivations. This is the case of the *Shomrei Adamah* (Guardians of the Earth) movement, which seeks to create greater awareness of contemporary environmental problems, especially through education (Tirosh-Samuelson 2002).

Islam

Islam is the second most practiced religion in the world after Christianity, with just over 1.2 billion followers. Although traditionally associated with the Arab world, most Muslims actually live in other geographical areas, with Indonesia, Pakistan, India and Bangladesh being the countries with the largest number of adherents. Nevertheless, Islam certainly has a strong Arab cultural root, which derives from the environment in which its founder, Muhammad (570–632), was born and lived. Original from Mecca and belonging to one of the most deeply rooted clans in the city, he worked as a merchant for a wealthy widowed merchant, Jadiya, whom he married. In 610 he received a revelation that would be the origin of the Koran, compiled decades after his death. He began to preach a strict monotheism, which confronted him with his own tribe and most of his countrymen, who guarded the Kaaba, then the base of the polytheistic cult. Tensions in Mecca led him to move to Medina in 622 (the beginning of the Muslim calendar). From there he consolidated his position and extended his dominion to other cities, conquering Mecca in 630, dying two years later. Endowed with a brilliant personality, Muhammad was able to unite the various Arab tribes and initiate a period of expansion that would be consolidated by his successors, especially Omar, from 638, after the conquest of Jerusalem (638).

There are currently two main branches of Islam: the Sunni (*Sunna* = Tradition), which includes 85% of Muslims, and the Shiite, which includes 13%. The division between the two began with the dispute over the legitimate succession of Muhammad, between the supporters of Ali (cousin and son-in-law of Muhammad) and those of the Umayyads (supporters of maintaining the succession to the original tribe of the Prophet). The former, the Shiites, lost the battle of Siffin (657) and were reduced to the regions where they were predominant, being now the majority in Iran, a good part of Iraq, Yemen, Lebanon and Syria. In practice, both branches of Islam have substantial differences in the interpretation of the Koran and liturgical practices.

Although Muslims do not worship Muhammad, he is considered the prophet *par excellence*, who received and transmitted God's ultimate revelation to humanity. This revelation is included in the Koran, the holy book of Islam. Of lesser importance, but also relevant for Muslims are the *Hadiths* (words and events of Muhammad's life), which provide a better understanding of the meaning of the Koran and the Islamic legislation (*sharia*). The Koran was probably transcribed in the time of the third caliph, Uthman, in 651–652, about 20 years after Muhammad's death. It consists of 114 chapters (or Suras), organized according to their length, with a preface, which is a short prayer that forms an essential part of the Muslim liturgy: 'In the name of Allah, the Merciful, the Compassionate'.

The legal interpretation of the Koran is carried out in the various legal schools, which can have very varied approaches, since Islam has no central authority (except for Shiism, which has imams, authorized interpreters of the Koran) (Morales 2001). The main schools are:

Hanafite, quite liberal, very widespread in the past and now mainly in the Balkans and the Caucasus.

Malikite, more rigorous, North Africa and some countries of the Persian Gulf.

Shafiite, predominant in Egypt, Yemen, Somalia, Eritrea and SE Asia.

Hanbalite, not tolerant of other religions, predominant in Saudi Arabia and Qatar.

Although most Islamic countries are governed by their own civil law, the *Shari'a plays* an important moral role, especially in the more rigorous schools. In addition, there are other sapiential interpretations of the Koran, collected in commentaries by mystics or sages in different traditions.

The religious life of the Muslim is organized around five pillars: the *Shahadah*, the profession of faith, is the most frequent act among Muslims: 'I confess that there is no God but Allah and that Muhammad is the Messenger of God'; the ritual prayer, highly regulated, not only in timetables, but also in content and places (mosques or in clean natural sites); almsgiving, *zakat*, initially 10% of earnings, although there are now different interpretations; fasting during Ramadan, the month in which the Koran was revealed, which involves abstinence from all food and drink, as well as sexual relations, during the hours of daylight; and the pilgrimage to Mecca (*Hajj*), once in a lifetime, which also follows very specific rites (Morales 2001).

In terms of its relationship with environmental conservation, Islam was born in an ecologically fragile environment, the Middle East, especially in terms of water resources. Perhaps for this reason, the Koran contains references to the good use of the Earth's resources. Like other monotheistic religions, it firmly believes in the existence of a Creator of all that exists. External nature has an objective and ontological existence (Ozdemir 2003), it has been created by God and, therefore, manifests His power and goodness, at the same time that it is a visible means for us to know some of his virtues. At the same time, the beauty and diversity of creation should lead us to express gratitude to God for his creatures. The contemplation of nature brings man closer to the divine, because he is present everywhere. It is like a sacred book, full of symbols, a theater in which the signs of God are manifested. The Earth belongs to God and praises Him, it is a living organism. Everything created has a specific purpose, it has been ordained by God with a specific meaning; there is nothing superfluous or the fruit of chance:

> Verily, in the creation of the heavens and the earth, and in the succession of night and day there are indeed messages for all those endowed with insight, (and) who remember God, standing, sitting and when they lie down and meditate (thus) on the creation of the heavens and the earth: O our Sustainer! Thou didst not create (any of) this without meaning and purpose. Infinite is Thy glory! Preserve us from the punishment of the Fire!
>
> (Qur'an, Surah 3, 190)

Thus, every creature has an intrinsic dignity, rights, and they serve not only human needs, but also the needs of other creatures.

Other texts, emphasize the pre-eminent role of humans in nature and enjoin them to thank God for granting them the creatures for their use. In the current context of Islam, this apparent contradiction between dominion and care of Creation is resolved in favor of granting humans a vicarious role in managing the Earth's resources, similar to the concept of environmental stewardship, common to Judaism and Christianity. Indeed, for Ibrahim Ozdemir (2003) the Qur'an suggests that human beings are God's vicegerents on Earth, are charged with caring for the planet's resources and, consequently, will be judged for their actions related to the natural environment. This kind of environmental responsibility can be considered parallel to that of one's own family, and God will hold us accountable for both:

> O men, adore your Lord who has created you and those who were before you, and fear God, who has made the earth a carpet for you and of the sky a castle, and has made water come down from the sky with which to extract from the earth those fruits that are your daily food.
>
> (Qur'an, Sura 2, 21:22)

In the same vein, Nasr (2003) argues that human beings are both servants (*abd*) and vicegerents (*caliph*) of God for the rest of Creation. Human beings can administer the natural goods if they maintain the condition of servants by being loyal to God's dispositions over the other creatures. Other Muslim authors disagree with this concept, as it perpetuates an anthropocentric and monarchical view of the relationship between humans and the natural environment (Afrasiabi 2003), in parallel with some criticisms that have emerged in the Christian world about the concept of environmental stewardship.

For Islam, animals also deserve special attention from God and therefore should be given special attention by humans. They have been created by God as such and not only to serve man, so an animal cannot be killed without a justifiable reason. They also glorify God:

> The seven heavens, the earth and all that is in them glorify Him. There is nothing that does not glorify Him with praise, but ye cannot perceive their glorifications.
>
> (Qur'an, Sura 17:44)

Animals also have an eschatological meaning, they often appear in the images of paradise, and some authors interpret that there they will have the ability to glorify God. They give names to several chapters of the Koran: the cow, the ox, the bee, the ant, the spider. In good part they are symbols of other spiritual realities, but they are also analogies close to the environment in which Muhammad originally preached.

God creates the world with a certain plan, with a balance:

> (By His command) the sun and the moon follow their appointed courses; (before Him) the stars and the trees prostrate themselves. He hath raised the heavens, and established a measure (for all things), that ye (also, O men) may never exceed the measure (of right): weigh ye then (your deeds) with equity and fall short of the measure!
>
> (Qur'an Sura 55: 5–9)

This balance should also govern man's relations with other human beings and with the natural environment around him, leading to a moderate use of resources. Some *hadiths* collect sayings of the Prophet that condemn the wasteful use of resources. In this sense, the symbolic importance of water, which in the Muslim world serves as an element of inner purification (following the Jewish and Christian tradition), is highlighted in this sense. All mosques have several fountains at the entrance for ritual ablutions, which are essential for prayer. Even though it is a ritual resource, some *hadiths* show the importance of using it with moderation. Ozdemir (2003) collects one of these passages, in which a messenger of God sternly admonishes a worshipper for wasting water in the ablutions. The reproved one wonders:

> Can there be waste when performing ablutions?' to which Allah's messenger replies, 'Yes, even if you perform them on the banks of a flowing river'.

For Ozdemir this text illustrates well a key principle of the Islamic consideration of the environment, since if even in the sacred use of resources (ablutions), wastefulness attitude is severely reprimanded, even in the case of unlimited resources (flowing river), how much more will it be in the most superfluous use of other goods of the Earth. The waste of resources causes pollution and loss of balance in the natural environment, affects living beings, violates the rights of future generations to those resources, is arbitrary and is not directed at satisfying basic needs, and goes against God's original design.

The current reality of human-environment relations in the Islamic world seems to contradict these principles markedly. At least that is the view of Nasr, one of the Muslim scholars who has studied the subject most thoroughly and who has pioneered environmental awareness from an Islamic perspective:

> When one studies the Islamic view of nature and society's relationship with the environment, when one analyzes how classical Islamic civilization created a society, and especially an urban environment, in harmony with nature, and compares it with the current situation in the Islamic world, it is obvious to conclude that neither Muslim governments nor most of their people are following Islamic principles in their dealings with the natural environment.
>
> (Nasr 2003, 87)

According to this author, the current disconnect between Islam and the environment would be caused by several factors:

Uncritical acceptance of the scientific and technological approach coming from the West, which is fundamentally secular and involves losing the sacred sense of nature and our relationship with it, in favor of a reductionist materialism. This condescending attitude affects not only the most pro-Western governments, but even the most fundamentalist ones, and is quite widespread among the intellectual leaders of these countries.

The migration of the rural populations to large cities has led to a loss of daily contact with the natural environment. The great masses of the new urban population, almost always impoverished, have no connection with the surrounding environment, which is otherwise completely artificial and very degraded.

Solutions to contemporary problems are sought in ways of approaching life that are alien to Islamic traditions, rather than in one's own cultural tradition. There is, to a large extent, a certain inferiority complex in the face of Western innovations.

Many Islamic countries are in the hands of dictatorial or authoritarian governments, which limit or eliminate the internal capacity to criticize development projects that cause significant environmental. In this sense, there is a lack of intermediate social groups to spread environmental awareness.

Environmental issues are not a priority for religious leaders, who place more emphasis on social issues and the personal relationship with God, as well as on preserving their own identity in the face of cultural confrontation with the West.

In this sense, Nasr suggests that environmental issues be part of education, both in universities and in the popular schools where many of these leaders are trained. For other authors who are more critical of Islam's environmental perspective, the lack of Islamic leadership on environmental issues is also due to a defensive position in the face of what many Islamic intellectuals see as a cultural invasion by the West (Afrasiabi 2003). For example, in the case of Iran, with a strong mix between political and religious components, no relevant ayatollah has addressed environmental issues in his magisterium, and references to the natural world are explicitly anthropological, for example, regarding controversies about pure and impure animals for human food.

Mawil Izzi Dien (2003) reviews other controversial aspects of environmental management in Muslim countries that have led to remarkable imbalances. For example, the enormous influx of resources in the Persian Gulf and in Saudi Arabia countries itself has led to the creation of clearly unsustainable cities and agricultural enterprises, in marked contrast to the dictates of the prophet on the reasonable use of water (in Riyadh, for example, water consumption is about 412 liters per person per day, compared to the 27 liters consumed by an average family in rural areas). Along the same lines of water management, conflicts over the shared use of river basins are important, for example, the Tigris and Euphrates between Turkey, Iraq and Syria, or the Nile between Egypt and Sudan. Dien (2003) points to the possibility that some of these international conflicts could be resolved through an environmental application of Islamic law.

Christianity

In the Dock

According to different authors, the main cause of the environmental crisis is the arrogant and selfish attitude towards other species that leads humans to use the Earth for their exclusive benefit. The origin of this attitude would be in the self-proclamation of human beings as masters of the planet, supported by their growing capacity to transform it through the immense potential offered by our ingenuity. For some thinkers, this human arrogance has a religious origin; it would obey a certain cosmology that confers on human beings a central role in Creation, superior to all other beings, who would have been created for their use and enjoyment. The logical consequence of this mentality is an attitude of despotic domination, which would be ultimately responsible for the ecological imbalances we observe today.

To which religious mentality is this accusation referring? Mainly to the Judeo-Christian tradition, which, by placing the human being in a central role, would have disconnected him from his natural origins. At the origin of this hypothesis is the article published by the historian Lynn White Jr. in 1967 in the journal *Science*, reflecting his intervention at the annual congress of the American Association for the Advancement of Science. Two elements justify, in the author's opinion, the accusation that there is a direct relationship between Christianity and the ecological crisis. On the one hand, the anthropological conception of Christianity, which places the human being at the center of Creation, above all other creatures, and, on the other, the great advance of science occurring in Christian Europe, which gave Western countries the technical capacity to transform nature for our own benefit.

White bases his accusation on the role attributed to human beings in the Bible, which gives them a unique role as the only creatures made in the image and likeness of God, which, according to White, implies a sharp separation and a marked pre-eminence of humanity over the rest of the creatures. He cited the first chapter of Genesis (1:28), which gives men and women the task of filling the Earth and having dominion over it. In White's view, this cosmological vision of the Judeo-Christian tradition implies a relationship of complete subordination to nature unknown to other religious traditions:

> Christianity, in absolute contrast to ancient paganism and the religions of Asia (with the exception of Zoroastrianism), not only established a dualism between nature and man, but also insisted that it is God's will that man exploit nature for his own ends.
>
> (White 1967: 1205)

This attitude, again following White, was strengthened by Christianity's determination to suppress the pagan religions it encountered in the course of its expansion, both in the European regions, where it initially spread, and in the territories it later settled. Primitive paganism often involved the worship of natural elements, such as certain places, animals or plants, which was a form of idolatry, severely rejected in the Bible.

In short, for White, Christianity is dualistic, separates humans from nature, and makes humans think of themselves as above other creatures, considering them only for their own utility. In contrast to this dominant tradition in Western Christianity, White highlights the figure of St. Francis of Assisi, who he believes is a remarkable example of a cosmic humility that places humans on a level of equality with other created creatures. St. Francis' approach is so innovative in relation to the anthropological conception of Christianity at that time that, according to White's interpretation, it is surprising that his message was not rejected by the Catholic hierarchy:

> The first miracle of St. Francis is the fact that he did not end his days at the stake, as many of his more radical followers would later do. He was so clearly heretical that a General of the Franciscan order, St. Bonaventure, a great and kindly Christian, tried to suppress the first vestiges of Franciscanism.
>
> (White, 1967, 1206)

The second factor linking Christianity to the ecological crisis, according to White, comes from the historical connection between scientific development and Christianity. This fact has been explained by different authors based on several reasons (Brooke 1991; Woods 2005). On the one hand, Christianity proposes a 'reasonable' God, who designs the Universe with stable, intelligible rules which can be apprehended by human beings, and which also bring them closer to a greater knowledge of God. In other words, science was considered as a means to better understand the creation and then the Creator, as an extension of natural theology that tries to understand better God through the study of creatures. On the other hand, in Christian Europe, the first universities were founded in the 12th and 13th centuries, which allowed the accumulation of classical and, later, oriental knowledge, towards the scientific emergence of the 16th and 17th centuries, with prominent figures such as Copernicus, Galileo, Newton, Descartes, Kepler, Pascal, to name only the most relevant scientists. Added to this was technological development, which allowed improvements in agricultural production, the emergence of craft industries, and the development of navigation, which became more and more reliable and circumnavigated the globe. From the 17th century onwards, scientific development and technology converged, giving Western countries an unprecedented capacity to transform the natural environment. Science was practiced by intellectuals, contemplatives, while technique was in the hands of manual workers, craftsmen, guided more by observation and tradition than by knowledge of causes. From the convergence of the two, tools began to become available to transform nature on a scale and intensity previously unknown: means of transportation, hydraulic works, large-scale agricultural enterprises, urban growth, etc.

Since this nature-dominating religious approach is at the root of the planet's imbalance, the solution to environmental problems for White is not technological, but religious: 'Since the roots of our problem are primarily religious, the remedy must also be essentially religious, whether we call it that or not' (White 1967, 1207).

This author does not go so far as to recommend the repudiation of Christianity in favor of other religions, but rather a reflection that allows us to change the anthropocentric vision in favor of a more balanced conception with the environment, precisely following the proposal of St. Francis, whom he proposes as the patron saint of ecologists.

So much for Lynn White's position, which has served to stimulate a debate of enormous impact. His article is cited in many academic papers,[3] which tend to put Christianity in the dock. It is striking that the accusations are made precisely in virtue of some characteristics of Christianity that, until a few decades ago, were considered symptoms of progress, such as the overcoming of paganism (which was considered a form of superstition, of worship of material elements that prevented material development), of humanism (which assumes the centrality of the human being over other creatures) or of its commitment to the rationality of knowledge (which allowed the birth of modern science).

Although White's article has been the most controversial, it was not the first text to accuse the Judeo-Christian tradition of having stimulated the environmental degradation of the planet. This same thesis was implicit in the forward of the main book of Aldo Leopold, in which he stated that:

> Conservation is getting nowhere because it is incompatible with our Abrahamic concept of land. We abuse land because we regard it as a commodity belonging to us. When we see land as a community to which we belong, we may begin to use it with love and respect.
>
> (Leopold 1949, viii)

Although he does not develop the underlying idea, this passage from Leopold is in line with advocating a new moral-ethical relationship with the natural environment, which would have to be given by a change in our cosmological vision, overcoming the anthropocentrism that would come from a biblical vision of human creation (what he calls the 'Abrahamic conception').

Response to Criticism

The provocative thesis of White's text has given rise to an extensive debate in which his arguments have been answered from various viewpoints. This has made it possible to enrich the environmental orientation of Christianity and to better explain the theological basis and practical implications of Christian approach to conservationism.

The defense against White's charges can be summarized in two blocks: on the one hand, those authors who try to demonstrate that there is a Christian tradition of environmental conservation that is not limited to the figure of St. Francis, and, on the other hand, those who aim to clarify the genuine exegesis of the biblical texts quoted by White.

With regard to the exceptional character of the life and message of St. Francis of Assisi (1182–1226), and in particular of what we could call his 'ecological theology', various authors have shown that he is certainly an extraordinary figure, but not an exclusive one in the history of Christianity (Binde 2001; Delio et al. 2008; Kim 2017; Sorrell 1988). St. Francis is not completely detached from the Christian tradition, he is a continuator – outstanding, no doubt — of a legacy that starts almost from the first Christians. The first reference of the Judeo-Christian tradition on the relationship of man with creation would be the biblical texts, which clearly reflect the sense of praise to God of all creatures, and therefore that they all have a purpose in themselves (to give glory to God) independent of human needs, as we have already mentioned in allusion to Judaism, with which Christianity shares tradition and sacred texts (the first part of the Bible, being called the 'Old Testament'). It is interesting to note that these are not marginal quotations, but texts of great relevance in Christianity. In fact, some of these passages of Sacred Scripture have been used daily by many religious communities for centuries, since they belong to the Liturgy of the Hours (the official prayer of the Catholic Church). This would be the case of the Hymn of the three young men, taken from the book of the prophet Daniel, which shows in a very vivid way how the adoration of God is not limited to human beings, but reaches out to the entire created universe:

> All you creatures of the Lord, bless the Lord, extol him with hymns forever and ever. Ye angels of the Lord, bless the Lord; ye heavens, bless the Lord. Ye waters of space, bless the Lord; ye hosts of the Lord, bless the Lord. Sun and moon, bless the Lord; stars of heaven, bless the Lord. Rain and dew, bless ye the Lord; all ye winds, bless ye the Lord. Fire and heat, bless the Lord; cold and frost, bless the Lord. Dews and snowfalls, bless the Lord; icebergs and ices, bless the Lord. Frost and snow, bless the Lord; night and day, bless the Lord. Light and darkness, bless the Lord; lightning and clouds, bless the Lord. Let the earth bless the Lord, praise him with hymns forever and ever. Mountains and peaks, bless the Lord; all that sprouts in the earth, bless the Lord. O ye springs, bless ye the Lord; ye seas and rivers, bless ye the Lord. Cetaceans and fish, bless the Lord; birds of the air, bless the Lord. O ye beasts and cattle, bless the Lord; extol him with hymns for ever and ever. O ye sons of men, bless the Lord; bless the Lord, O Israel.
>
> (Daniel, 3:57–88)

Logically, for Christianity the obligatory reference in any doctrinal question is the life and preaching of Jesus Christ. The Gospels and the rest of the New Testament writings are not very explicit in this matter, probably because they take for granted a much kinder relationship with the environment than the present one, which they therefore do not need to counter. In any case, the contemplative closeness of Jesus to the land he traveled through appears in numerous passages. For instance, his prayer took place in natural places: on the mountain (Mt 14:23; Mk 6:45; Lk 6:12; Lk 9:28), in the desert (Mt 4:1; Mk 1:3; Lk 4:1), in the Garden of Olives (Mt 26:36; Mk 14:32; Lk 22:39). On the other hand, the preaching of Jesus Christ often uses natural images, showing the closeness of his listeners to these images. He uses animals as analogies for deeper ideas ('Be wise as serpents and harmless as doves', Matthew 10:16), or as a reference ('Look at the lilies, how they neither spin nor weave. But I say to you that not even Solomon in all his glory was clothed like one of them', Luke 12:27), or as an example ('To what shall we compare the kingdom of God, or by what parable shall we expound it? It is like a mustard seed which, when sown in the ground, is smaller than any seed sown on earth' Mark 4:30–31). Finally, Jesus also shows a detailed knowledge of the territory he inhabits: 'When you see a cloud rising in the west, at once you say, "It is going to rain", and so it happens. And when the south blows, you say, "It will be hot", and so it is' (Luke, 12:54–55).

The tradition of the first centuries of Christianity continues this closeness to the nature. On the one hand, there is a growing appreciation for the eremitical life, first in solitude and then in small communities that are the origin of the first monasteries. There are many testimonies of the so-called *desert fathers and mothers* (St. Anthony Abbot, St. Macarius, St. Pachomius, St. Thecla...), who appreciated the life in direct contact with the natural environment. The first monasteries founded in Egypt in the 3rd and 4th centuries, some of which still survive (such as St. Catherine or St. Anthony) amid a rather hostile natural environment, are based on this tradition. Often the places inhabited by monastic communities have sites of great environmental value, which have been preserved in an exemplary manner thanks to these communities (Mallarach et al. 2014). Some of the most representative examples of this tradition of environmental stewardship in Christian monastic communities are the holy mountain of Montserrat (Catholic, Spain) and the monastic republic of Athos (Orthodox, Greece).

For the first Christian theologians, the natural environment was considered the framework of creation, nature -as we understand it today- was a concept foreign to their reflection. Material creation is a manifestation of divine beauty and goodness; therefore, it is an icon -image- of the Creator. The contemplative tradition of Christianity, which is already explicit in the writings of the early Church Fathers, is framed within this meaning. For example, St. Augustine, who was in close contact with the classical poets, whom he knew well, gives free rein to his aesthetic sense in one of the last chapters of *The City of God*:

> What shall I say of the beauty, so great and so diverse, of the sky, of the earth and of the sea; of so great an abundance and so admirable a beauty of the light itself in the sun, the moon and the stars; of the freshness and thickness of the forests, of the colors and odors of the flowers, of so great a diversity and multitude of birds so talkative and painted, of the variety of species and figures of so many and so great animals, among which those that have less grandeur and body cause us the greatest admiration? And what shall I say of the beautiful spectacle of the sea when it is dressed as if in a livery of different colors, varying its color in many ways, from a red green to a blue green?
>
> (Augustine 413–426)

Other similar testimonies can be found in the lives of St. Martin of Tours, of St. Columbanus, or of St. Basil, author of the principal monastic rule of the East, who dedicates a series of sermons to the commentary on the first chapters of Genesis (Hexameron). The following paragraph is taken from this text:

> ...if sometimes, on a clear night, while watching with careful eyes the inexpressible beauty of the stars, you have thought of the Creator of all things; if you have ever wondered who has sprinkled the sky with such flowers, and why visible things are even more useful than beautiful; if sometimes, during the day, you have studied the wonders of light; if you have risen from invisible things to the invisible Being, then you are a good spectator, and are prepared to have a seat in this august and blessed amphitheater.
>
> (Basil 361–364)

The first Benedictine monks placed great value on manual labor as a complement to prayer, which led them to a close interaction with nature. The monks always lived in the same monastery, which required them to take care of the natural resources they lived on, practicing what we would now call sustainable agriculture and animal husbandry. The Cistercian reform in the 12th century, also showed several common views with St. Francis in its appreciation of beautiful places, and in its sense of the praise of God from all creatures. There are many writings in this line, deserving to be highlighted *'The description of Clairvaux'*, of great literary beauty. In another writing, in the form of a dialogue, one of the first Cistercians, Idung of Prüfening (1153–1174), includes the following admonition between two monks:

> You are a very simple theologian if you do not recognize that inanimate creatures, devoid of meaning and life, speak to God and give glory to God. What else would be the meaning of the Hymn of the young and the praises of the Psalms, where all creatures are invited to praise God?
>
> (Quoted by Sorrell 1988, 31)

Also outstanding is the figure of St. Hildegard of Bingen (1098–1178), German Benedictine, a woman of great culture, who had a remarkable influence on her time. In her poetic works she also showed an enormous knowledge of nature and a genuine inspiration for the beauty of creation:

> As the Creator loves his creation, so the creation loves the Creator. Creation, naturally, was designed to be adorned, to be shown, to be gifted with the love of the Creator. The whole world has been embraced by this kiss.
>
> (quoted by McDonagh 1986)

To sum up, following Sorrell (1988), we can affirm that the roots of Franciscan spirituality are nourished by various sources coming from the primitive Church. It gathers a hermit tradition and, at the same time, is linked to the classical poetic tradition, from Ovid to Virgil, which expanded in the Middle Ages in the songs of the troubadours, whom the Saint of Assisi knew well in his youth. There is also a certain scientific tradition, collected in the compilations of classical knowledge of St. Isidore, St. Bede or Rabano Mauro, which give a very characteristic accent on the observation to nature. Finally, it is also nourished by a communitarian tradition of the first Benedictine and Cistercian monasteries, which showed a contemplative attitude and care for the environment around them. It is not for nothing that some of the Christian monasteries can be considered among the most resilient inhabited places in the world, with many centuries of uninterrupted occupation.

In this tradition, the figure of St. Francis of Assisi (1182–1226) emerges in a very particular way, not only because of his relationship with other creatures, but also because of other aspects of his spirituality, which makes him deserve a central role in the history of Christianity. As far as his closeness to nature is concerned, the spiritual approach of St. Francis certainly represents a leap over the preceding tradition. Starting from a profound piety and a hermit tradition of austerity and solitude, the life of St. Francis denotes multiple manifestations of an extraordinary union and affection for all creatures. His biographers interpret this closeness to wild animals as a manifestation of his innocence and holiness of life, as a reflection of a harmony that goes back to the state before the original sin.

> For him, the whole of nature was a kind of temple of the divinity, where the Creator could be discovered, contemplated and served. The whole universe becomes a visible sacrament of the presence of God and the action of Christ.
>
> (Merino 2008, 38)

His sermon to the birds marks a unique moment in his life, a certain interior conversion, which involves extending his apostolic vocation to all creatures: in short, including them also in the preaching of the Gospel, as a manifestation of family affection, of care and service. The deepest attitude of St. Francis is that of a natural mysticism, which leads him to connect directly with God in the contemplation of creatures, to transcend the Creator from the created. This interior attitude is extraordinarily reflected in the song of Brother Sun, one of the most beautiful poems of medieval literature:

> Most High, omnipotent, good Lord, thine are the praises, the glory and the honor and every blessing. To you alone, Most High, they belong, and no man is worthy to make mention of you. Praise be to you, my Lord, with all your creatures, especially for the lord brother sun, which is day, and by which you enlighten us. And he is beautiful and radiant with great splendor, from you, Most High, he bears significance. Praise be to you, my Lord, for sister moon and stars, in the sky you have formed them bright and precious and beautiful. Praise be to you, my Lord, for

> brother wind, and for the air and the cloudy and the serene and all weather, by which you give sustenance to your creatures. Praise be to you, my Lord, for sister water, which is very useful and humble and precious and chaste. Praise be to you, my Lord, for brother fire, by which you lighten the night, and he is beautiful and joyful and sturdy and strong. Praise be to you, my Lord, for our sister mother earth, who sustains and governs us, and produces various fruits with colorful flowers and grass. Praise be to you, my Lord, for those who forgive for your love, and endure sickness and tribulation. Blessed are those who endure them in peace, for by you, Most High, they shall be crowned. Praise be to you, my Lord, for our sister bodily death, from which no living man can escape. Woe to those who die in mortal sin: blessed are they whom I find in your most holy will, for the second death will not harm them. Load and bless my Lord, and give him thanks and serve him with great humility.
>
> (Saint Francis 1225)

Various authors have discussed the ultimate meaning of the preposition 'for' in the canticle, which St. Francis uses in the first stanzas. For some, it indicates the one who is asked to perform the action, which would imply the saint's desire that the various natural elements give glory to God. According to this interpretation, this text would not be very different from others that we have already cited, such as the song of the three young men, where the creatures are also asked to give glory to God. A second sense, much more innovative, gives the 'for' the meaning of 'because of', which would imply that the saint is asking people to praise God in gratitude for the marvel of beings he has created. In this second sense, the Canticle perfectly reflects the Franciscan message, which logically should not be seen as an ecological plea (we are in the 13th century, let us not forget), but as a new appreciation of creation, which accompanies man in his prayer, but also helps him to raise his spirit to God, as a manifestation of divine goodness and beauty. His approach, then, departs from a dualistic vision that was evident in other Christian movements considered heretical by the Church, such as the Cathars, contemporaries of St. Francis, who, taking up the Gnostic and Manichean tradition, considered precisely the material as repugnant and advocated a spiritualist asceticism far from *material contamination*. It is significant that the saint of Assisi is also attributed with the tradition of plastically reconstructing Christmas scenes (the popular 'Christmas crib'), since it shows an appreciation for the material environment that accompanied the birth of Jesus. For St. Francis, this material dimension is also part of the divine plan of creation and, therefore, should not only not distance us from God, but rather should bring us closer to his presence, elevating us from the contemplation of creatures to the perfections of their Creator. It is a kind of natural mysticism, where the natural serves as a framework for our encounter with God.

The spirituality of St. Francis has had a great impact ever since. His radical way of life was, and still is, difficult to generalize, but this does not imply that his message was ever considered suspect of heresy, as White suggested in his article. At no time in his life was the orthodoxy of St. Francis, who was canonized in record time, just two years after his death, questioned. Nor is it true that St. Bonaventure mutilated the most controversial aspects of his biography, since his work on the saint is essentially the same as what we have from other truly reliable sources, as the most profound experts on his work have shown:

> No Catholic biographer, whether earlier or later, has ever considered the Franciscan interpretation of the environment as heterodox. Moreover, in discussions of Franciscan affairs in the thirteenth century, either with papal advisors or with the popes themselves, the subject never came up in the discussions.
>
> (Sorrell 1988)

The second group of responses to White's criticisms of Christianity came from biblical exegetes, who suggest giving a less anthropocentric explanation of the biblical passages than the one assumed by the American historian (Conradie 2006; Habel 2011). Of particular interest is the commentary on the first chapter of Genesis, where the most solid argument for the environmental accusation against Christianity is based. Here is the passage alluded to:

> Be fruitful and multiply and replenish the earth and subdue it; command the fish of the sea and the birds of the air and every creeping thing that creeps upon the earth.
>
> (Gen 1:28)

The literal interpretation of this verse would certainly give rise to an anthropocentric cosmology, where God would grant humans the ability to exploit the Earth for their own benefit. However, as in many other texts of Sacred Scripture, the Christian tradition considers the literal interpretation to be one of the possible interpretations, and not necessarily the most appropriate in certain passages. As old as the school of Alexandria (Origen, for example, in the 3rd century), one of the most important in the first centuries of Christianity, was the conviction that there are multiple meanings in the sacred texts, and therefore the need to interpret them in the light of other passages to extract their ultimate meaning, as well as to respect the tradition (as it has been previously interpreted) and the own exegetical authority of the Church.

In this sense, if the first chapter of Genesis is read in the light of other texts, the supposed divine 'dominion' over the rest of creation is very nuanced. As is well known, Genesis includes two accounts of Creation, in its chapters 1 and 2. Gen. 1 describes the evolutionary cycles of Creation and the role of the various creatures in the divine plan, showing the beauty and goodness of the whole ('and he saw that everything was good' in several verses of this chapter), while Gen. 2 describes a different account of the origins, focusing on the garden of Eden, the role of the first human (Adam) and his relationships with other animals and then the first female. Most exegetes now believe that the first two chapters come from different Jewish traditions: Genesis 1 and 2:1–3, commonly called priestly, and Genesis 2:4–25, now called non-Priestly (Carr 2020), which is about 500 years older than the former. In this second account, man is presented as the fruit of the Earth, extracted from the same material to which the rest of the creatures belong, which in a way marks his fragility ('For you are dust and to dust you shall return', says the same book of Genesis in the following chapter (3:19). The name *Adam* comes from *Adamah*, which means soil, earth. The Latin has maintained this same meaning, since the term human and *humus* have the same root; in short, we are made of the same material as the rest of Creation, although the sacred text certainly indicates that we are the fruit of a special intervention of God, who breathes into us the breath of life.

After the creation of man, Genesis 2 indicates that Yahweh took man and placed him in the Garden of Eden 'to till it and to keep it' (Gen 2:15). The image of Eden, then, is not that of a hunting ground, but of a garden given to man by God to be properly tended. In short, creation was not given by God to man as a sovereign patrimony, but was entrusted to his care, which implies its responsible use. This implies that the dominion referred to in Genesis 1, insofar as this dominion was given before original sin when man and woman were in harmony with the rest of creation, is not to be understood as an absolute license, but rather as a delegated stewardship.

On the other hand, the term dominion has a different meaning in the Jewish people of biblical times than the one we use today. The chosen people were governed by prophets (Moses, Joshua) in the first instance, then by Judges (Othniel, Gideon, Deborah, Samson...) and only five hundred years after the exodus from Egypt the monarchy is proposed ('set us a king to judge us, like all the nations', request the elders of Israel in the first book of Samuel: 8, 5). Significantly, Holy Scripture indicates that this request displeases God, who grants it as the fruit of the obstinacy of men. Moreover, the first kings (Saul, David, Solomon) are anointed by a prophet, since they are really chosen by God to lead his people; therefore, they are not absolute lords, but also delegates of divine authority, as in fact happens with Saul, from whom God takes away the throne because of his disobedience (1 Samuel 15:10–23). This allows us to understand better the term 'rule the earth' in the context of vicarious rule, which requires responsibility on the part of the one who exercises it before the one who commissions it.

Custodians or Priests of Creation

This interpretation of how the Bible understands the relationship between human beings and the rest of creation is the basis for the concept of *environmental stewardship*, which is currently the most widespread paradigm among

Christians with environmental concerns, although it can also be linked to Judaism and Islam. It is based on the premise that human environmental behavior must be accountable to God for our use of the planet, since the resources of creation do not belong to us but rather have been delegated to us by God. In this sense, we are stewards of the rest of creation, responsible for the fulfillment of the natural purpose for which they were created. In this sense, environmental care is not only an ethical responsibility, but also a religious one. It is also a recognition that if all creatures exist because of God's will, then all have intrinsic value in God's eyes.

The biblical image of the steward of other people's resources comes from some parables of the Gospel, such as the unfaithful steward (Luke, 16), the steward of the vineyard who pays the workers (Matthew, 20: 8), the lord who gives talents to his employees to negotiate in his absence (Matthew, 25: 14–28), or the tenant of the vineyard who has to give the fruits at the proper time (Matthew, 21: 33–41; Luke, 13: 6–9). Jesus Christ praises the 'faithful and wise steward', whom the Lord puts 'in charge of all his property' (Luke, 12: 42–46, and Matthew, 24:45–51). Jesus Christ is certainly not using this parable to refer to the management of natural resources, but rather to illustrate to his listeners the moral dependence of one's actions: they are accountable to God. Now, this delegated responsibility can also be applied to environmental issues. In a nutshell, it means acknowledging that the world was created by God, who is the Lord of the universe, and that this world has been entrusted by God to human beings made in His image. This makes us stewards, managers, or custodians, even if we do not realize it. It entails careful stewardship of the planet's resources, not selfish exploitation. Moreover, it must consider not only the present inhabitants of the planet, but also the future ones, and remember that the world has a value in itself, regardless of its usefulness to human beings.

For some authors, although this concept represents a more benign interpretation of the anthropocentric vision of Christianity, it is still based on a mistaken approach to the relationship between human beings and the natural environment. For example, for Clare Palmer (2006) the concept of environmental stewardship is associated with the medieval idea of feudal dependence: the serf is entrusted by the lord, who is the absolute owner of the goods over which he delegates. Aside from the fact that this figure implies a social regime that has already been superseded and has despotic connotations that are not acceptable today, Palmer asks 'in whose name do we exercise this administration?'. The concept is present in the New Testament, as we have indicated, but is not explicitly applied to the stewardship of the Earth. On the other hand, the management of goods requires sufficient knowledge of how they work and how they relate to other goods, and we do not know enough about the Earth system to manage it properly. Nor is it clear that this human management is necessary for the proper functioning of the planet, since in fact the Earth has been functioning without our help for many millions of years, and it is not clear that our intervention has always perfected it. From a more theological point of view, Palmer points out that the concept of stewardship could indicate that God is absent from the world, as if He were incapable of exercising this dominion by Himself, which would call into question the divine providence over all things. In the same vein, other authors have suggested that the idea of environmental stewardship implies an image of God as an absentee landlord who overrides the spontaneity of human life on Earth.

The supporters of the concept of environmental stewardship do not consider these criticisms to be sufficiently solid, since the figure of delegated dominion is present in many passages of the Bible, which reinforce God's sole ownership of Creation, as we have already seen in the texts of Leviticus. On the other hand, the 'cession for use' that God grants to human beings does not imply an abandonment of his ordinary providence or an elimination of human freedom, just as the freedom of the tenant of a property or the director of an investment fund does not disappear; both manage other people's resources, and to that extent they have to deliver a balance of their management, but according to their own decision in order to make them perform more adequately. Therefore, the concept of God's delegated stewardship would lead us to be more responsible in the use of the planet's resources, directing them to satisfy our vital needs or those that make us great as persons, instead of doing so despotically or capriciously, in addition to providing for the good of other creatures, in accordance with God's original design when creating them. It would be a matter of being co-creators, if we can speak in this way, of bringing to fulfillment the purpose for which God created all the beings that inhabit this planet, of collaborating with Him in His care for creation, as a Christian scientist beautifully suggests:

> God is committed to his creation. He eloquently demonstrated this by sending his son Jesus to be part of creation and giving us the responsibility to be good stewards of creation. More importantly, we do this not on our own, but in partnership with him, a partnership that is so beautifully presented in the early chapters of Genesis, where we read that God walked with Adam and Eve in the garden at sundown.
>
> (Houghton 2006, 317)[4]

The orthodox Christian tradition has developed a parallel idea to stewardship, but with a stronger theological background. John Zizioulas (2013) proposed in this line the concept of being 'priest of creation', which he considers to be better suited to the Christian worldview than the stewardship model, since the latter implies a managerial and conservationist approach to nature, and thus a utilitarian and mechanistic view of creation. In addition, Zizioulas believes that it implies a passive concept that reduces human action to preservation rather than development. In this opinion, the concept of priest of creation provides a more positive vision of our relationship with nature, considering that humans are the only material creature with freedom and creativity, with a spiritual dimension, and therefore the only capable of offering material realities to God, spiritualizing them: 'only the human being is united with creation while being able to transcend it through freedom' (Zizioulas 2013, 167). We are the only species capable of uniting the world and relating it to God for its own benefit (not just for ours). Christ is the perfect Priest of Creation, since by His Death and Resurrection He restored the unity and communion of all creation with God. Following Zizioulas, the understanding of the human beings as priests of creation implies an existential, rather than functional (as in the steward model), relationship with nature, because it affects what one is, rather than what one does, and because the destruction of nature has not only moral (it is wrong) but also ontological (it ceases to be) consequences:

> As priest of creation we have the unique mission and immense responsibility of uniting God and the material world. Our task is not simply to preserve creation but to purify it and elevate it to the level of divine existence.
>
> (Zizioulas 2021, 58)

If caring means sustaining and supporting the development of other creatures, this 'ecological priesthood' implies an even greater responsibility: to strive for the full perfection of the natural world, to bring it back to God (with the goal of sanctifying the world), but also to restore its material well-being, to heal the effects of human sin (not only original sin, but ongoing transgressions that affect the integrity of the web of life) on nature.

In the Old Testament, Noah is the paradigmatic example of the human vocation to be a priest of Creation, as he builds an ark where all creatures find salvation from the Flood, but also where he cares for them all. This requires a docile attitude to do God's will, but also humility to care for the specific needs of each creature during the days of incessant rain. Both care and humility are important components of the priestly soul, which bases an interest in building bridges to others who may be far removed from God in a spirit of service and generosity:

> Noah's ability to humble himself to the demands and needs of the animals. Yet these characteristics are precisely what led to Noah's being proclaimed righteous. Noah is righteous because like God the creator he took upon himself the maintenance of all creation.
>
> (Wirzba 2003, 33)

In short, the religious concept of custodians and priest of Creation implies that human beings have a responsibility before God for the care of Creation, for maintaining the legacy we have received. This implies that caring for the natural environment is part of our religious righteousness, and that it is therefore a moral obligation to avoid its superfluous use, as well as to repair the damage that human activity may have caused.

Environmental Empathy

Another complementary approach to human-nature relations in the Christian tradition holds that our moral actions have consequences for the natural environment. In this conception, material nature is linked to human destiny, to the decisions we freely make, accepting their consequences, whether positive or negative. In this vision, which we can call 'environmental empathy', the human being would be like the moral representative of creation, which rejoices or suffers in solidarity with human moral deeds. The idea of environmental empathy has been used by other scholars to refer to the impact of nature in our feelings and affections, following the traditional definition of empathy as 'the understanding and sharing of another person's emotional experience' (Davis 1983), but also to what extent this empathetic attitude affects environmental concern (Scultz 2002). In this case, we use the term referring to the moral bond that theologically exists between humanity, the only material creature with freedom to commit moral evils and goods, and the rest of material creatures (Chuvieco 2012, 2024).

This idea is reflected in various passages of the Bible, where it is evident that the moral goodness or badness of man has repercussions on the well-being of the environment: the earth is fertile when the chosen people are faithful to Yahweh, while infertility and desolation are the consequence of their sins. The most obvious passage is that of the story of the original sin itself, when the soil loses its natural fertility as a consequence of Adam and Eve's disobedience.

> To the man he said: 'Because you listened to your wife and ate from the tree of which I had forbidden you to eat, Cursed be the ground because of you! In toil shall you eat its yield all the days of your life Thorns and thistles shall it bring forth to you, as you eat of the plants of the field. By the sweat of your face shall you get bread to eat, Until you return to the ground, from which you were taken; For you are dirt, and to dirt you shall return'.
>
> (Genesis 3:17–19)

The infertility of the earth is linked to man's wrongdoings; the loss of ecological balance is linked to the loss of moral balance, so that human's sin brings with it the degradation of nature: 'I will destroy the earth, and your enemies will be astonished when they come to occupy it' (Leviticus 26:32). The Earth withers because of man's unfaithfulness to the covenant with God, as is recorded in a text of the prophet Hosea:

> There is no fidelity, no mercy, no knowledge of God in the land. False swearing, lying, murder, stealing and adultery! in their lawlessness, bloodshed follows bloodshed. Therefore, the land mourns, and everything that dwells in it languishes: The beasts of the field, the birds of the air, and even the fish of the sea perish.
>
> (Hosea 4:1–3)

Therefore, recovering this balance demands people's conversion:

> If I close heaven so that there is no rain, if I command the locust to devour the land, if I send pestilence among my people, and if my people, upon whom my name has been pronounced, humble themselves and pray, and seek my presence and turn from their evil ways, I will hear them from heaven and pardon their sins and revive their land.
>
> (2 Chronicles 17:13–14)

If this is so, then the final regeneration of man, with the Second Coming of the Messiah and the Last Judgment, will also mean the complete restoration of the natural balance. In this sense, we find very interesting a text of the prophet Isaiah, traditionally interpreted in a messianic tone. In fact, it is a common liturgical reading in the weeks leading up to Christmas:

> The spirit of the Lord shall rest upon him, the spirit of wisdom and understanding, the spirit of counsel and might, the spirit of knowledge and the fear of God; the fear of the Lord shall inspire him (...Then) The wolf shall dwell with the lamb, the panther shall lie down with the kid, the calf and the young lion shall eat together, and a little one shall watch over them. The cow shall graze with the bear, her young shall lie down together; the lion shall eat straw, like the ox, the sucking child shall play by the hide of the snake, the newly weaned shall put his hand into the serpent's den. No one shall do evil or harm in all my holy mountain, for the earth is full of the knowledge of the Lord as the waters cover the sea.
>
> (Isaiah 11:1–10)

Here it is announced that there will be a new time, in which the Messiah will restore the original harmony, which affects not only the human being, but also the relations between animals. Finally, in this line we can also include a controversial passage from the letter of St. Paul to the Romans, which seems to indicate that the material creation needs the conversion of man for its full realization:

> Creation, in fact, was subjected to vanity, not spontaneously, but by him who subjected it, in the hope of being freed from the bondage of corruption to participate in the glorious freedom of the children of God. For we know that the whole creation groans to this day and suffers the

> pains of childbirth. And not only it, but we too, who are the first fruits of the Spirit, groan within ourselves, longing for the ransom of our body.
>
> (Romans 8:19–23)

The Catholic Church and Environmental Conservation

As we have seen, there is a rich environmental tradition in the early centuries of Christianity, which was consolidated in the Middle Ages through the land management practices of some monastic orders (Benedictines, Cistercians) and the influence of Franciscan spirituality. However, environmental issues have not been addressed in Catholic Church documents until relatively recently, perhaps because it has not been a controversial theological topic.

The first references to environmental problems appear in the social encyclicals of John XXIII and Paul VI, mainly in *Octogesima Adveniens* (1971), although it was John Paul II who dealt with this topic in greater detail in some of his social encyclicals: *Solicitudo Rei Socialis* (1987) and *Centesimus annus* (1991) and, above all, in his message for the World Day of Peace in 1990 (1990). He developed a broad pastoral approach to the ecological question, which fits well with his passion for the contemplation of nature and mountaineering, along with his personalist philosophy, which unites natural and human ecology.[5] In his social encyclicals he analyzed the ecological question within the framework of the integral development of the most disadvantaged peoples, stimulating the concept of environmental stewardship. Following the Catholic tradition, John Paul II insisted that the Holy Scriptures justify the use of natural resources to serve the needs of human beings, but do not approve of their waste and degradation, which does not correspond to God's original mandate, but would rather be a consequence of the imbalance caused by original sin:

> Made in the image and likeness of God, Adam and Eve were to have exercised their dominion over the earth (Gen 1:28) with wisdom and love. Instead, they destroyed the existing harmony by deliberately going against the Creator's plan, that is, by choosing to sin. This resulted not only in man's alienation from himself, in death and fratricide, but also in the earth's 'rebellion' against him (cf. Gen 3:17–19; 4:12).
>
> (John Paul II 1990, 3)

This rupture with the original equilibrium not only affected us, but also our relationship with the environment, which is now hostile and does not facilitate our maintenance except after hard work. Consequently, the tension with nature has a moral origin. Following the idea of environmental empathy that we mentioned earlier, John Paul II stated that,

> When man turns his back on the Creator's plan, he provokes a disorder which has inevitable repercussions on the rest of the created order. If man is not at peace with God, then earth itself cannot be at peace.
>
> (John Paul II 1990, 5)

In other words, the ecological problem is caused by the loss of the original balance of man with God and, consequently, of man with the natural environment. Thus, the remedy should not only be technological, but primarily moral. In this context, John Paul II mentioned the possibilities of genetic manipulation, which allows the development of new species of plants and forms of animal life that can have a significant impact on the natural ecosystem, as well as on human health. He also advocated for respecting the integrity and original balance of the cosmos, recalling that the Earth is a common heritage, 'whose fruits should be for the benefit of all', rejecting the selfish appropriation of resources by a few. He also affirmed the need to promote a new international solidarity to help solve the food and environmental problems of emerging countries, which are so often the cause of armed conflicts. He insisted on the need to move towards a more sober way of life that would guarantee the maintenance of the planet's habitability and access to resources for the most disadvantaged:

> Modern society will find no solution to the ecological problem unless it takes a serious look at its lifestyle. In many parts of the world society is given to instant gratification and consumerism while remaining indifferent to the damage which they cause. As I have already stated, the seriousness of the ecological issue lays bare the depth of man's moral crisis. If an appreciation of the value of the human person and of human life is lacking, we will also lose interest in others and in the earth itself. Simplicity, moderation and discipline, as well as a spirit of sacrifice, must become a part of everyday life, lest all suffer the negative consequences of the careless habits of a few.
>
> (John Paul II 1990, 13)

Benedict XVI reflect on the ecological question in his encyclical *Caritas in veritate* (2009), as well as in the message for the celebration of various World Days of Peace (2008), and in multiple speeches on his apostolic journeys (perhaps the most significant being the one he addressed to the German Parliament in 2011).[6] In his message for the 2010 World Day of Peace, he showed its serious moral implications:

> Can we remain indifferent before the problems associated with such realities as climate change, desertification, the deterioration and loss of productivity in vast agricultural areas, the pollution of rivers and aquifers, the loss of biodiversity, the increase of natural catastrophes and

> the deforestation of equatorial and tropical regions? Can we disregard the growing phenomenon of 'environmental refugees', people who are forced by the degradation of their natural habitat to forsake it – and often their possessions as well – in order to face the dangers and uncertainties of forced displacement? Can we remain impassive in the face of actual and potential conflicts involving access to natural resources? All these are issues with a profound impact on the exercise of human rights, such as the right to life, food, health and development.
>
> (Benedict XVI 2010, 4)

For Benedict XVI, there is a clear connection between the environmental crisis and the preservation of world peace. On the celebration of the 2007 World Day of Peace, he pointed out that

> The destruction of the environment, its improper or selfish use, and the violent hoarding of the earth's resources cause grievances, conflicts and wars, precisely because they are the consequences of an inhumane concept of development. Indeed, if development were limited to the technical-economic aspect, obscuring the moral-religious dimension, it would not be an integral human development, but a one-sided distortion which would end up by unleashing man's destructive capacities.
>
> (Benedict XVI 2007, 9)

One of the pillars of this ecological conversion would be inter- and intra-generational solidarity. We do not have the right to irresponsibly consume the resources of entire societies that need them for their own development (the poorest countries, often rich in resources that are exploited by others), nor do we have the right to deprive from future generations, to whom we must leave an environmental legacy that will allow them to develop satisfactorily. He recalled that the universal destination of goods, confirmed by the Second Vatican Council, must be the foundation of intergenerational solidarity:

> Natural resources should be used in such a way that immediate benefits do not have a negative impact on living creatures, human and not, present and future; that the protection of private property does not conflict with the universal destination of goods; that human activity does not compromise the fruitfulness of the earth, for the benefit of people now and in the future.
>
> (Benedict XVI 2007, 8)

Since every economic decision has a moral character, Benedict XVI suggested that economic activity should also take into account environmental costs, in such a way as to assess their possible loss or deterioration, incorporating mechanisms to restore them as far as possible, or at least mitigating their effective impact.

> When making use of natural resources, we should be concerned for their protection and consider the cost entailed – environmentally and socially – as an essential part of the overall expenses incurred.
>
> (Benedict XVI 2007, 7)

Many examples can be cited in this area, from the restoration costs already included in public works, to fines for industrial discharges, or the recently proposed taxes on greenhouse gas emissions. There is no doubt that the recognition that natural resources are not infinite helps to place an economic value on them, which in turn makes it easier to conserve them. He considered that the need for an economic assessment of the impact of infrastructures on nature is particularly evident in the case of energy production, since all forms of energy have environmental costs: aesthetic (wind energy), atmospheric (fossil fuels), surface (solar energy), health (possibly nuclear energy), etc.

> To be sure, among the basic problems which the international community has to address is that of energy resources and the development of joint and sustainable strategies to satisfy the energy needs of the present and future generations. This means that technologically advanced societies must be prepared to encourage more sober lifestyles, while reducing their energy consumption and improving its efficiency.
>
> (Benedict XVI 2007, 9)

One of the main contributions of Benedict XVI to the ecological debate was the emphasis on linking human and natural ecology. He considered that respect for the natural balance affects equally human nature as well as other living beings. In his last encyclical, *Caritas in veritate*, he insisted on this idea, defending:

> In order to protect nature, (…) the decisive issue is the overall moral tenor of society. If there is a lack of respect for the right to life and to a natural death, if human conception, gestation and birth are made artificial, if human embryos are sacrificed to research, the conscience of society ends up losing the concept of human ecology and, along with it, that of environmental ecology. It is contradictory to insist that future generations respect the natural environment when our educational systems and laws do not help them to respect themselves. The book of nature is one and indivisible: it takes in not only the environment but also life, sexuality, marriage, the family, social relations: in a word, integral human development. Our duties towards the environment are linked to our duties towards the human person, considered in himself and in relation to others. It would be wrong to uphold one set of duties while trampling on the other. Herein lies a grave contradiction in our mentality and practice today: one which demeans the person, disrupts the environment and damages society.
>
> (Benedict XVI 2009, 51)

Finally, Pope Francis has dedicated very diverse references to the ecological question in his texts, beginning with the homily with which he began his pontificate (Pope Francis 2013b), although the most significant is undoubtedly the publication of his encyclical dedicated specifically to this question, *Laudato Si'*, in 2015 (Pope Francis 2015). The enormous moral authority of his Magisterium and the example of his simple and sober life ensure that this is a text with great impact, both for Catholics and for other people with environmental sensitivity, Christians or not.

In the homily that opened his pontificate, he took the figure of St. Joseph (whose feast day was celebrated that day) as an example of the custodian of the Holy Family and encouraged the faithful to be custodians of others and of all creation:

> The vocation of being a 'protector', however, is not just something involving us Christians alone; it also has a prior dimension which is simply human, involving everyone. It means protecting all creation, the beauty of the created world, as the Book of Genesis tells us and as Saint Francis of Assisi showed us. It means respecting each of God's creatures and respecting the environment in which we live.
>
> (Francis 2013b)

In the same homily, in the context of an attitude of care, of loving kindness towards all people and all creatures, the Pope asked the powerful of the world for a greater commitment to the task of environmental stewardship:

> Please, I would like to ask all those who have positions of responsibility in economic, political and social life, and all men and women of goodwill: let us be 'protectors' of creation, protectors of God's plan inscribed in nature, protectors of one another and of the environment.
>
> (Francis 2013b)

For Pope Francis, care for creation is an expression of appreciation for God's gifts to humanity and of responsibility for the use we make of this treasure, in order to bequeath it in its entirety to future generations.

The main document of Pope Francis on the ecological question is the encyclical *Laudatio Si'* (LS from now on), dedicated exclusively to this topic. The title takes up some words from the *Canticle of Brother Sun* of St. Francis of Assisi, whose spirituality illuminates many of the passages of the text. It is a text that goes beyond ecology to offer a profound reflection on our model of development and the impacts it has on people and nature. It is a beautifully written text, which has offer both great praise and some criticism, paradoxically from people close to Catholicism. Some have pointed out that it is an attack on modernity. Certainly, it is an attack on modernity born of enlightened rationalism and the industrial revolution, which considers nature as a storehouse and human beings simply as instruments of production. Throughout the encyclical, the denunciation of environmental and social degradation is closely linked, because both are the fruit of the same selfish attitude.

> The human environment and the natural environment deteriorate together; we cannot adequately combat environmental degradation unless we attend to causes related to human and social degradation.
>
> (LS, n. 48)

The encyclical has the significant subtitle 'On Care for the Common Home', and throughout its pages emphasizes the connection between environmental and social degradation, which reflects the personal motivation of Francis towards environmental issues, as in developing countries social crises are intermixed.[7] The text brings together the voices of many different bishops' Conferences, from both industrialized and developing countries, who speak of their daily experience of environmental problems in their nations: it is not a theoretical text, prepared by those who study the problems in the solitude of a laboratory, but is very close to the concerns and needs of those who feel the effects of environmental degradation on a daily basis. The encyclical also includes texts by other Christian leaders, such as the Patriarch of Constantinople, Bartholomew I, who has done a great work to raise environmental awareness in recent decades.

At several points in the encyclical, the Pope points out the moral implications of the ecological question, encouraging Catholics to make a greater commitment to the environment. In contrast to the position of those who believe that this is a minor issue, he points out:

> It must be said that some committed and prayerful Christians, with the excuse of realism and pragmatism, tend to ridicule expressions of concern for the environment. Others are passive; they choose not to change their habits and thus become inconsistent. (…). Living our vocation to be protectors of God's handiwork is essential to a life of virtue; it is not an optional or a secondary aspect of our Christian experience.
>
> (LS, n. 217)

The first part of the encyclical is aimed at understanding the problems and fragility of the 'common home'. The review of the most relevant topics includes climate, water, loss of biodiversity, oceans, energy, local pollution, urban inequalities, and even the 'spiritual' pollution caused by the digital world... Since the beginning of our presence on Earth, there has always been human intervention in nature, but adapting to the conditions it offered. However, when humans have ignored these conditions, they have moved from being a part of nature to being in confrontation with it. Faced with this situation we find in society two types of reactions: those who cling to a 'technological optimism' and those who think that man's presence must be limited for the good of nature. The Pope denounces both extremes. Technology dominates society so much that life is objectified and reduced, making the environmental and human problem irrelevant. There is a false technological optimism and a tendency to consider that the laws of the market will correct imbalances. He invites us to change our way of thinking because of the unsustainability it entails and warns us:

> The pace of consumption, waste and environmental change has so stretched the planet's capacity that our contemporary lifestyle, unsustainable as it is, can only precipitate catastrophe.
>
> (LS, n. 161)

The necessary change cannot be limited to superficial aspects; it requires profound changes in our relationship with the environment, in our use of resources, but also in our personal and social goals. Hence the Pope's far-reaching challenge: 'it is simply a matter of redefining progress. Put simply, it is a matter of redefining our notion of progress'. (LS, n. 194).

On the other hand, Francis reminds us that we are also part of the natural balance, that it makes no sense to build an ecology apart from people:

> A sense of deep communion with the rest of nature cannot be real if our hearts lack tenderness, compassion and concern for our fellow human beings. It is clearly inconsistent to combat trafficking in endangered species while remaining completely indifferent to human trafficking, unconcerned about the poor, or undertaking to destroy another human being deemed unwanted. This compromises the very meaning of our struggle for the sake of the environment.
>
> (LS, n. 91)

The second part of the encyclical extends the theological arguments to care for nature. Genesis shows us how human existence is a relationship: with God, with our neighbor and with the Earth. Human's sin has broken and denaturalized this relationship, transforming it into conflict. It is therefore necessary to recover the awareness of this divine presence present in every creature, because everything is destined to the final restoration in God:

> The ultimate purpose of other creatures is not to be found in us. Rather, all creatures are moving forward with us and through us towards a common point of arrival, which is God, (…) Human beings, endowed with intelligence and love, and drawn by the fullness of Christ, are called to lead all creatures back to their Creator.
>
> (LS, n. 83)

The Pope shows a concrete example of what he proposes as 'ecological conversion' in the figure of St. Francis, who encountered God through the contemplation of His creatures. Nature is Creation: it is a gift given by a Father who bears a promise of Love. The life of man and of every creature is not the fruit of chance or chaos, but each person -and ultimately each creature- has been willed by a Creator God, who gives everything created a meaning of its own. Beyond human interest, all beings have been freely created by God, so they have an intrinsic value in His eyes, beyond any instrumental value they may have for us or for other creatures.

This has practical consequences in our daily actions, in our consumption habits, in the way we use energy, in the way we transport ourselves. The Pope suggests some areas where we can concretize this kinder relationship with the environment,

> such as avoiding the use of plastic and paper, reducing water consumption, separating refuse, cooking only what can reasonably be consumed, showing care for other living beings, using public transport or car-pooling, planting trees, turning off unnecessary lights.
>
> (LS, n.211)

These are only examples. The encyclical is not a catalog of good practices, nor can it be reduced to something as superficial as throwing waste in the right bin. The question goes deeper. In Francis' view, the ecological question:

> cannot be reduced to a series of urgent and partial responses to the immediate problems of pollution, environmental decay and the depletion of natural resources. There needs to be a distinctive way of looking at things, a way of thinking, policies, an educational program, a lifestyle and a spirituality which together generate resistance to the assault of the technocratic paradigm.

In short, the ecological conversion proposed by the Pope in this encyclical goes beyond establishing new habits.[8] It is a matter of approaching the world as a gift from a God who loves us, of recognizing His presence in all created things, of contemplating them as a means of knowing Him better and accompanying them in their original purpose. Along with this new perception, it is also necessary to change our scale of values to analyze what is superfluous in our lives, what harms the Earth and affects other human beings and other creatures. This requires a moral rearmament, emphasizing the goods that enhance us as human beings, instead of maintaining an absurd rhythm of consumption. We should also remind ourselves more often that happiness is not tied to the possession of material goods. The economy cannot be the only criterion of choice. It is only a means. In short, Francis invites us to break with personal selfishness, to think more about others, those of today and those who will come after us, to enjoy the beauty of Creation and to thank God for it.

> If we approach nature and the environment without this openness to awe and wonder, if we no longer speak the language of fraternity and beauty in

> our relationship with the world, our attitude will be that of masters, consumers, ruthless exploiters, unable to set limits on their immediate needs.
> (LS, n. 11)

The encyclical ends with a hopeful tone, encouraging the readers to overcome social and personal selfishness, as not everything depends on us, on our weaknesses, but also on a God who keeps us in his providence, together with other creatures. The Pope ends by imploring God for this change with two very beautiful prayers, which unite believers of other spiritual traditions and other Christians, because prayer is a vital part of this ecological conversion.

After the publication of LS many other initiatives from the Vatican, and personally from Francis, have continued reinforcing the main message of the text. For instance in October 2023, he released a new document focused on climate change, the apostolic exhortation 'Laudate Deum', with the occasion of the feast of St. Francis, in preparation of the COP28 in Dubai. The document emphasized the relevance of climate change and encouraged world leaders to take more ambitions measurements to mitigate its negative effects. Faced with the potentially catastrophic consequences of this environmental problem, Pope Francis encourages all believers to avoiding skeptical postures and changing the consumerist mentality. He proposed strengthening global agreements to reduce GHG emissions, with concrete and binding measures that also consider compensation for the most affected and vulnerable countries.

Synthesis: Religious Values and Environmental Conservation

In the summary we have made of the different religious approaches to environmental conservation, it is clear that there are many aspects common to all of them, which allow us to foresee a convergence of the major religions in their response to the ecological crisis. Certainly, none of the religious traditions has as its primary goal to provide solutions to the environmental problem. Nor is any religion truly anthropocentric, even though they all start from principles that govern relationships with other human beings, since they all focus on the relationship between humanity and God. In short, all religions are theocentric, not anthropocentric or biocentric, since they are oriented to cultivate our spiritual dimension. In short, any religion tries to explain the great questions that we all ask ourselves at some point in our lives: what is the meaning of our life, where do we come from, what happens after we die, what principles should guide our relationships with others?

Although for no religious creed is it a priority to regulate human's relationship with nature, it is obvious that one's religious approach will have a significant influence on one's relationship with the environment. Growing environmental sensitivity, on the one hand, and greater awareness of the magnitude of the environmental problems we face, on the other, have led us to rethink human-environment relations from different perspectives, both philosophical and religious. These new approaches can significantly help us to face the solutions to the crisis with a much richer perspective, going beyond the mere application of technology to be based on ethical and moral principles that will make our relationship with nature much more harmonious and stable in the future.

Summarizing what we have reviewed in the previous sections, we can group the religious approaches to human-media relations into six main approaches (Chuvieco 2012), which we summarize in Figure 6.2.

Dominion: Nature is at the service of human beings, with the only limitation being that the use of resources must be sustainable, i.e., that we do not cause serious harm to the future inhabitants of the planet. This approach could be considered as an absolute dominion of man[9] over creation, with the only restriction being the search for the common good. Some literal interpretations of the first chapter of Genesis could support this approach, although Islam also considers man as superior to the rest of Creation and the rest of the religions in some way prioritize the most pressing needs of men over the rest of the species.

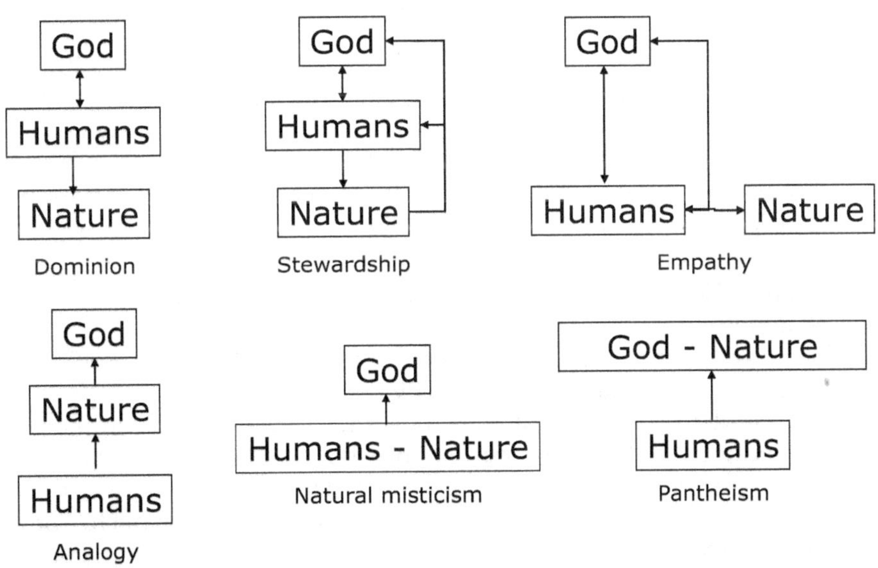

FIGURE 6.2
Summary of religious traditions in the God-human-nature relationship. (Adapted from Chuvieco 2012).

Environmental Stewardship: Human dominion is exercised in a manner delegated by God, which implies that we are responsible to God Himself for the environmental impact of our actions. In this approach, humans would be a delegated steward of the goods of creation. God delegates to us a responsible stewardship of the other species that inhabit the planet, caring for their preservation, not only for their continued use in the future, but also for the dignity of these creatures themselves, created by God for a specific purpose. This approach would include chapter 2 of Genesis, which applies to both Christians and Jews, as well as the concept of vice-caliphs of creation found in Islam.

Environmental empathy: This approach, quite common not only to monotheistic religions, but also to Hinduism and Buddhism, recognizes a certain moral communion between man and the rest of creation. When we behave properly, the Earth yields its fruits, while the Earth is hostile when human behavior is morally wrong. Man's responsibility is to accompany the rest of creation in its prayer, orienting his life towards God so that other creatures may also accompany him in this praise.

Analogy: This attitude implies considering nature as a manifestation/revelation of God, which completes the written revelation (of sacred books, such as the Bible) through His works. Man discovers God through the perfections of the world around him, which can be seen as an image of the goodness, greatness, and beauty of God. This is the basis of the so-called natural theology, which is quite traditional in Christianity, but is also present in Islam or in various branches of Hinduism and in many indigenous traditions.

Natural Mysticism: A step further in this consideration of the world as an image of God would be to contemplate it as a path to intimate union with God. In the Christian tradition, the clearest example would be St. Francis of Assisi, without forgetting other later mystics, such as St. John of the Cross. In the Islamic tradition, this approach is present in some Sufi schools. In Hinduism, we see it reflected in various texts of the Vedas.

Pantheism: This religious tradition considers nature as a material representation of God, worshipping and adoring it. This is explicitly forbidden in monotheistic religions and is present in the paganism of some indigenous religions as well as, in some Hindu and Buddhist schools. This sacred consideration can be given by the analogy of God (incarnations of some deity), by remembrance of some sacred situation (some trees in Hinduism) or by the properly pantheistic consideration (God is part of all beings). In this case, it is a more generic worship, which usually takes place in certain natural places of special symbolic character for the community.

As we have seen, all major religions contain a balanced approach to human-environment relations that would make it possible to solve the planet's major environmental problems. As we have seen, several authors have argued that some religions are more environmentally friendly, while others, particularly Christianity, has been accused of giving religious support to an anthropocentric vision. If there were such marked differences in the environmental attitudes of the major religious traditions, it would seem consequential to find also significant differences in the state of environmental protection in the countries where these religious traditions are dominant.

This was the starting approach of a preliminary analysis we conducted a few years ago, selecting a set of countries that were sufficiently representative of the major religions (Christian, Muslim, Jewish, Buddhist and Hindu), and analyzing their most relevant environmental indicators (Chuvieco, et al. 2016). For this task, we selected those collected by Yale University in an index called EPI (*Environmental Performance* Index: http://epi.yale.edu/). The preliminary conclusions we reached did not allow us to conclude that Christian countries have lower levels of environmental protection than other countries with supposedly more environmentally friendly religious traditions (Chuvieco, et al. 2016). In other words, countries with a Buddhist or Hindu tradition not only did not show a better environmental situation than Christians (at similar levels of economic development), but on the contrary, they showed similar or even lower scores in some critical indicators. If countries with religions theoretically closer to environmental protection (such as Hinduism or Buddhism) do not offer better environmental quality scores than other supposedly more aggressive religions (such as Christianity), this could be explained by three reasons:

Because religions have little effective weight in shaping the values of a given society, even when they are in the social majority.

Because religions are not highly concerned with environmental conservation.

Because in reality Christianity is no more negative towards nature than other religions.

How do we measure the weight of religion in the social configuration of a country? It is always complex, but it seems obvious that there are multiple social differences between countries with different religious majorities, for example, in terms of food, work holidays or family customs. In fact, it can be shown statistically that, for instance, wine consumption is lower in Muslim-majority countries than in Christian ones, which we should associate with the Islamic prohibition of alcohol consumption.

If majority Christian, Muslim or Buddhist countries have different peculiar customs or social values, they should also have different environmental scores if they had different attitudes toward environmental protection. If this is not the case, then it seems that the most obvious conclusion seems to be that different religious traditions do not lead to drastic changes in environmental conservation practices in the countries where they are in the majority. Other explanatory factors for the state of environmental health would then have to be proposed. In our work we considered the level of wealth and the Human Development Index. Both were found to be directly related to environmental quality in Christian and Buddhist countries, but not in Muslim and Hindu countries, where negative relationships were even observed (Chuvieco, et al. 2016). Judaism (represented by the only Jewish state, Israel) also did not fare well in the analysis, with environmental quality indicators far below what would be expected for its level of wealth. In short, in some groups of countries it was observed that the constraint to conservation is not culture (religion) but economics (*per capita* income), while in others, even at high income levels, environmental conservation does not appear to be a national priority. The statement would be more clearly attributed to some Muslim countries, which have very high incomes combined with medium and low values in environmental indicators. In this sense, it would confirm some of the criticisms of the environmental priorities of Islamic countries that we discussed earlier.

From the analysis in this chapter, there is an interest in increasing the environmental awareness of religious leaders, as they could make a very relevant contribution to improving environmental practices, especially in societies with greater religious influence (Chuvieco, et al. 2016). Furthermore, it would be advisable to encourage reflection on the environmental implications of different religious cosmologies, while promoting cooperation between different religious traditions in environmental initiatives. It is worth recalling, however, that leaders of all the major religious traditions of humanity have encouraged their followers to make more committed decisions in solving environmental problems, through reaching interfaith declarations.[10] This is undoubtedly a very fertile front also for fostering interreligious dialogue, as the experience of the two interreligious meetings in Assisi (1987 and 2011) testified. We conclude this chapter with some particularly significant words in this regard, taken from the joint message of John Paul II and the Orthodox Patriarch Bartholomew I signed in June 2002:

> A growing awareness of the relationship between God and humanity brings with it a deeper sense of the importance of the relationship between human beings and the natural environment, which is God's creation and which God has entrusted to us to guard with wisdom and love.11

Exercise for Reflection

The professor can divide the classroom in 5–6 groups, each one being assigned to study and present the principles of different religions about the environment. The students may not be necessarily believers of that religious tradition. It is recommended to select the larger religious traditions, as they have more impact worldwide: Christianity, Islam, Judaism, Buddhism and Hinduism, but it can also be focused on differences within religious traditions (for instance, among the main Christian Churches: Catholic, Orthodox, Baptist, Lutheran, etc.).

Each group should document the main environmental position of that religious tradition, linking it to the main principles of that particular faith. They should select at least a key document that explains the environmental principles, either extracted from a sacred book of the religion or from the writings of an authoritative source (a relevant leader of that tradition). Present a critical analysis of those principles in comparison with the actual environmental actions promoted by that tradition, either in agreement or in contradiction with those principles. Reflect on the importance of religious convictions in those behaviors against other drivers of daily habits (social, legal, economic, etc.).

Once each group has presented its findings, a synthesis of the similarities and differences could be presented to the whole class in a debate where each group can present what it sees in its tradition as similar to others and how common environmental concerns can serve as a vehicle for interfaith dialogue.

It is important to avoid value judgments on the religious traditions per se: we should not be discussing here which tradition is more or less morally or theologically right: only the environmental consequences of their principles and whether they are being followed by the believers of that tradition.

Notes

1 Although it may seem anecdotal, we find it interesting to highlight the impact of the first film in the Avatar series, the highest grossing film to date, in denouncing this problem, since its central argument was precisely the usurpation of an indigenous territory by settlers who wanted to exploit raw materials. The new version of this saga (*Avatar: The Sense of Water*) extends the plot, now in the environment of another indigenous community living in a coastal region.

2 Quotations from the Bible come from the New American Bible, English translation of the Neo Vulgate by the United States Conference of Catholic Bishops, 2002.

3 Specifically, in 10,366 articles according to Google's academic search engine (last query, July 2024).

4 John T. Houghton (1931–2020) was an eminent English scientist, professor of atmospheric physics at Oxford University, director of the British Meteorological Centre and coordinator of the United Nations International Panel on Climate Change (IPCC) between 1988 and 2002.

5 For an extensive review on the main environmental themes developed by the Magisterium of John Paul II can be consulted Lorbiecki (2014).

6 See among others the works of Lindsey (2012) y Koenig-Bricker (2009).
7 In a recent book on post-pandemic (Papa Francisco 2020), among many other topics, Pope Francis takes stock of his concerns regarding the 'ecological question' and the drafting of Laudato Si'. He points out that this was not an instantaneous 'conversion' to these issues but was the fruit of contact with the witness of many people – especially the most vulnerable in poor countries – who made him rethink the relevance of the environmental question.
8 A brief reflection on the historical meaning of ecological conversion can be analyzed in Ormerod and Vanin (2016).
9 We are using man in these paragraphs and in the figure as embracing both male and female.
10 We were fortunate to organize a meeting with environmental scientists and religious leaders, where the interest of all religious traditions in fostering environmental engagement was evident (Chuvieco et al. 2016b).
11 Text available at https://vatican.va/content/john-paul-ii/es/speeches/2002/june/documents/hf_jp-ii_spe_20020610_venice-declaration.html.

7

Responses to the Ecological Crisis

Doing Nothing

One of the simplest, and most commonly useless, attitude towards solving problems is to ignore them. Our psychology sometimes tries to elude the need to take decisions that may affect our ordinary life by overseeing the problem, assuming that somehow it will be solved by others or, even worse, that it will disappear in the same way as it occurred. Health problems are an example of this attitude in some individuals, who assume that their bodies have mechanisms to heal external pathogens. This may be the case for some illness, but others certainly need support from additional interventions, either surgery or biochemical treatment. The opposite attitude is that of persons who visit the physician even at the smallest symptom of unexpected corporal behavior. In between these two extremes are most ordinary people, who visit the doctor when problems cannot be solved by ordinary remedies.

Bringing the same example to the health of the Earth system, there are individuals who consider that environmental problems do not exist, that unexpected observations are part of cyclical fluctuations. The case of climate change is paradigmatic, with many skeptics who, in spite of the growing scientific evidence of anomalies, still consider that there are not warming trends in the Earth climate, or they attribute that change to natural causes, most commonly without mentioning which are those natural factors or where their connections to observed datasets have been evidenced. These attitudes may be explained by different positions, from those that simply do not want to change their normal habits, which heavily rely on fossil fuels, to those that feel the problem is too global to make any significant impact from their particular contribution to those that simply consider that someone, nobody knows who and how, will solve the climate imbalance.

But as it occurs in many medical problems, the solution to climate and other environmental problems requires taking specific actions. Simply letting them go, ignoring the problems, is not going to solve them. Time does not cure everything. In environmental matters, doing nothing means maintaining processes that, according to our current scientific knowledge, are leading to a

DOI: 10.1201/9781003545972-7

level of alteration of the planetary system that could become extremely dangerous not only for other species (which are disappearing, as science strongly attests), but also for our own. We have already discussed the importance of the observed effects of climate change, deforestation or industrial activity. Some are local, with obvious processes of degradation, which may be irreversible. Others have a more global component and are therefore more difficult for ordinary citizens to observe but are potentially very serious.

Few climatologists any longer doubt that we are approaching tipping points in greenhouse gas concentrations or temperature that could lead to abrupt changes in the planet's climate. An elementary sense of prudence should lead us to much more drastic measures to change our energy model. Although the projected costs of the changes needed to mitigate the problem are very high, the costs of the projected impacts may be much higher. According to the report by the British economist Nicholas Stern (2007), the cost of implementing effective climate change mitigation measures could be estimated to be around 1% of global GDP, but the effect of doing nothing and making the temperature rise to 2–4° by the end of the century, could imply losses of 5 to 20% of global GDP. Considering the climate justice dimension of the problem, this economic burden cannot be assigned to those countries with less wealth, which in turn has the least contribution to the current accumulation of GHG. It is reasonable that the main cost of the energy transition should be financed by the developed countries, as they have more economic capacity and higher responsibility in the problem. The Paris Agreement covered this justice dimension and foresaw the creation of a $100 billion climate fund to help emerging economies in carrying out the energy transition, to base their development in renewable sources. Even if this amount seems huge, it is still very low compared with the immense challenge ahead. Emergencies like Covid-19 have shown us that it is possible to take drastic measures when the common good requires it. The economic losses from the pandemic are far greater than the economic effort required to change our energy model. It is possible, but we lack the will. Short-term interests, lack of leadership by those who should be making agreements, generational selfishness, can be very expensive. Passing the problem on to future generations is as unsupportive as enjoying a more luxurious home than we need and can afford, passing the mortgage payment on to our children (who may not even be able to use it).

Technology Will Fix It

We live in a society that highly values science and technology, in large part because of their contribution to improving the length and quality of our lives.

Never before have there been such astounding advances in our knowledge of the world around us, giving us the ability to transform it at will, to create new materials, new methods of production, cures for disease, transportation, or simply to expand our leisure activities. These advances in science and technology are generally viewed in a positive light, as if they are always for the benefit of humanity. Likewise, anyone who criticizes a scientific advance is considered at least suspicious, if not an ideological interloper, especially if the criticism is based on ethical issues. However, it would be naïve to deny the harmful effects of technology, such as the control of people (surveillance, information gathering, etc.), the manipulation of food or human embryos, not to mention nuclear, biological, or chemical weapons.

Science and technology are not intrinsically good or bad, but it would be foolish to think that all their consequences are always positive. Their natural limits are marked by knowledge gaps (what can be done and what is not yet feasible), or by the economic constraints of the moment (what is profitable and what is not yet profitable), or by the vested interests behind particular research projects. We should remember that science and technology are not always altruistically motivated. They are also funded to produce products that can be sold and ultimately enrich those who invent, design and manufacture them. It also serves as a tool to control, protect and, in the worst cases, destroy other human beings. In short, most of the great scientists agree on the need to establish ethical boundaries for science, recognizing that these boundaries are easily crossed when success or profit is sought at any cost. Francis Collins, one of the leaders of the Human Genome Project, said of bioethics committees:

> Scientists must play an important role, since they possess sufficient knowledge to distinguish clearly between what can and cannot be done. But scientists cannot be the only ones involved in such committees. Scientists by their very nature are hungry to explore the unknown. Their moral sense is generally no better or worse developed than that of other groups, but they are inescapably affected by a potential conflict of interest that may lead them to violate the boundaries set by non-scientists. Consequently, a wide variety of perspectives should be represented on such committees.
> (Collins 2006)

We can point to many recent examples of scientific advances that have proved controversial, such as embryonic stem cell research or attempts at human cloning. In this area of biomedical research, various ethics committees are trying to define the boundaries that are considered ethically inviolable. For example, the CRISPR technique, which is revolutionizing genetic engineering, has enormous potential to cure diseases, but also to manipulate the structure of living beings, with impacts that are still unknown. For this reason, the developers of this technique themselves (Francisco Juan Martínez Mojica,

Jennifer Doudna and Emmanuelle Charpentier, the last two winners of the Nobel Prize in Chemistry in 2020), have warned of the need to reflect on the ethical implications of this technique. For these reasons, we must emphasize that any technology should be judged not only by its effectiveness or its economic viability, but also by its capacity to make our lives worthier, happier, and more in line with our ultimate expectations as human beings.

The authors of this book are personally committed to the development of science and technology, which is part of our professional activity, but precisely because of this we are aware that science can be used for many purposes and can serve very different interests. We will not deny the evidence of the great progress that scientific and technological development has brought, but neither should we ignore the ethical implications that many of these developments bring with them, creating additional problems to those they seek to solve. As a general criterion, it seems obvious that it is better to act on the cause of a problem than on its effect, and only on the latter when the former cannot be solved. For example, if we observe that there is a leak in the roof, the sensible thing to do is to find the source of the leak and plug it: it is not enough to put a bucket on the floor so that the water does not damage the carpet.

As far as environmental conservation is concerned, technology can certainly help to solve some of the problems of deterioration and degradation that we observe, as well as contribute to the restoration of those that have already occurred, but we cannot entrust it exclusively with the solution of environmental problems. In some cases, we can even accuse it of having created them, such as those derived from nuclear energy (leaks, waste handling), pesticides and herbicides (impacts on fauna, flora and human health), transgenic crops (effects on biodiversity), or energy production (such as impacts caused by oil spills). Sometimes technology acts following economic interests that tend to take little account of the environmental impact it generates, especially if they seek to maximize short-term profit. In this context, technology can be a great ally of economic powers alien to the area where they are acting, to the detriment of natural processes and local populations. Suffice it to mention the impact of some of the major infrastructures that have been built in recent decades, such as the Three Gorges Dam in China, which displaced more than 1.5 million people.

In short, without underestimating the relevance of technology, we think that it does not address the root of the problem. Technology sometimes replaces a necessary change with a kind of *placebo*, a substitute that reduces the perception of the problem, but does not eliminate it. If there is erosion in a watershed due to deforestation in the headwaters and this causes the rivers to produce more floods, the solution is not to build check dams downstream, but to reforest upstream. More dams will mitigate the problems, but at a much higher cost because they are not natural solutions and because they create new problems (artificial breakage of flows, slope instability, etc.). In short, technology helps with ethical decisions, but does not replace them.

In the specific problem of climate change, technology will of course help us to discover energy sources with lower emission rates, to improve efficiency of renewable sources and to reduce the concentration GHG, among other things, but it would be very worrying if a technological solution to the problem were proposed without addressing its root cause, which is more moral than technical. For example, various geoengineering measures are being developed to achieve global cooling, such as the installation of atmospheric reflectors, spreading chemicals in the upper atmosphere to reduce incoming solar energy or extend ocean CO_2 absorption capacity by artificial fertilization. All these measures are undoubtedly cheaper than changing the current energy model, but what are the implications when so much is still unknown about the natural functioning of the Earth system? Perhaps a new glaciation, much more serious than the very warming that is being addressed. More worryingly, these impacts would be global, even if only a few countries decide to implement them, further exacerbating climate injustice (Dannenberg and Zitzelsberger 2019).

On the other hand, trusting the whole solution to technology implies a pride attitude, as if we had 'everything under control', and that it is only a matter of time before we achieve production, transportation or consumption systems that are less aggressive with the environment. Certainly, history shows us that technological development has objectively improved the environment in many places, but it is also true that it has worsened it in others, adding polluting elements foreign to nature that it is now impossible to assimilate. The most obvious case is that of the chlorinated compounds that were used in refrigeration, involving a serious deterioration of the stratospheric ozone layer or the extended impacts of pesticides (DDT). They solved one problem (maintenance of food or thermal comfort in the places where they were used), and created another more serious problem, putting at risk the natural filter of ultraviolet radiation, which is essential for life on the planet. Moreover, scientific and technological development allows us to modify the environment on a scale and with an intensity unprecedented in human history. Critical voices in this regard come from the field of experimental science itself. Erwin Chargaff, one of the most important chemists of the 20th century and the father of DNA, spent the last years of his life denouncing the excesses of microbiology and genetic engineering, which would have unpredictable consequences for the environment and our own lives. In a letter to the journal Science, he said:

> This world is given to us on loan. We come and we go; and after a time we leave earth and air and water to others who come after us. My generation, or perhaps the one preceding mine, has been the first to engage, under the leadership of the exact sciences, in a destructive colonial warfare against nature. The future will curse us for it.
>
> (Chargaff 1976, p. 940)

Indigenist Response

Environmental criticism of the current economic and social model is sometimes interpreted as a frontal rejection of contemporary society, which has corrupted both the environment and human beings. Consequently, one response to environmental problems would be a return to more primitive ways of life, or at least a move away from the environments where Western society manifests itself in a more radical way, as in the case of large cities. This is well expressed in the 2007 film, *Into the Wild,* which tells the story of a young American who flees his father's home in search of new experiences to fill his existence. His final destination is Alaska, considered an ecological paradise far from any contact with Western society. Based on a true story (Krakauer 1998), it ends tragically and makes the viewer reflect on the extent to which this existential journey is supported by a somewhat simplistic approach to what it means to live 'in and from' the natural environment, and what are the roots of personal and environmental problems.

On the other hand, it is also a somewhat simplistic generalization to assume that the less technologized a society is, the happier, healthier, and more in tune with the natural environment it will be. Perhaps, those who propose a kind of 'return to the Paleolithic', are re-editing the 'myth of the good savage', so much commented upon in previous centuries. While it is evident that indigenous cultures deserve all our respect and that we have much to learn from them, it is also true that their traditions are not always as peaceful and in harmony with the environment as some simplistic views claimed for. For example, the American anthropologist Jared Diamond, who lived for many years with tribes in New Guinea, shows that tribal societies can have warlike conflicts of similar or greater magnitude than industrial societies, logically taking into account as a comparison the proportion of the population affected by these conflicts (2013).

On the other hand, we must not lose sight of the fact that traditional societies have rather a fragile way of life, with a low life expectancy, as a result of an unbalanced diet and very elementary access to medical procedures. In fact, they tend to have high infant mortality rates, and although some people can live very long, the age of death is much lower than that observed in Western society.

Indigenism as an ecological ideal underlies some of the criticisms that have been voiced with some frequency about the environmental impact of European colonization. For example, Nasr considers that the reverential attitude to nature of the Native Americans

> made possible the preservation of an entire continent in a state of almost Edenic perfection until the arrival of the Europeans, which entailed the accelerated destruction of the natural environment.
>
> (Nasr 2006, 106)

Perhaps there is a certain Manichaeism underlies in this paragraph, which contrasts the most primitive cultures, which supposedly would have achieved a perfect balance with nature, with Western civilization, which would only be oriented towards the abusive exploitation of the environment. Certainly, there was environmental destruction with the arrival of Europeans to the American continent, but it would be naïve to deny the existence of environmental problems before that colonization. On the one hand, several examples have been demonstrated of migrations (or even extinction) of primitive peoples due to resource depletion caused by overexploitation, as in the case of the crisis of the Mayan civilization or, more locally, of the inhabitants of Easter Island. On the other hand, examples of widespread extinction of species linked to the migration of primitive peoples have been documented, as in the case of the colonization of New Zealand by the Maori in the 12th–13th centuries A.D. (Bowman et al. 2009), or the extinctions of large animals caused by the entry of hunting tribes into America some 14,000 years ago (Barnosky et al. 2004).

In short, proposing a return to the Paleolithic as an alternative to the current social model seems neither possible nor reasonable. We have much to learn from traditional civilizations, but we can also help them to overcome some of their vulnerabilities. Certainly, these cultures are much closer to the natural environment, which may serve us as inspiration, but they also have ways of life that would be impossible to generalize to the population currently inhabiting the planet.

Are We Too Many?

Some of the most critical thinkers about environmental concern and action indicate that the ecological thought has an anti-humanist bias. By emphasizing that human beings are the main cause of environmental problems, if humanity is the 'cancer of the planet', a solution then is to eliminate human beings. Population control, eugenics and deep-ecology movements join forces in this attitude. They imply that we are 'too many', but they usually do not state who is responsible for 'being too many', nor by what criteria those exceeding the 'assumed limit' should be chosen and eventually eliminated.

These position, of course, have very controversial points: on the one hand, what is the maximum volume of human population that will allow us to maintain a balanced coexistence with the rest of the living beings and, on the other hand, what to do with the 'surplus' between this limit and the population that currently inhabits the planet (in 2023, just over 8 billion people). Neither of these two aspects is easy to clarify. Several authors have given some figures to estimate what they call the 'carrying capacity' of the planet, in terms of human population. The well-known biologist Paul Ehrlich stated a few decades ago that the Earth cannot support more than 2 billion people. Ehrlich was an entomologist by training, but he published several books on population issues. The most famous was *The Population Bomb*, published in 1968. In this book, he predicted a whole series of catastrophes as a consequence of population growth. All of the predictions made in this book have not been fulfilled, as their assumptions were incorrect, but it is still one of the most cited texts in the field of demography, still receiving awards that are not confirmed by his ability to predict future scenarios.

Other authors, even without mentioning a population limit, have insisted on the delicate balance between population growth and resources, concluding that we have exceeded the earth's ideal population limit and that it is necessary to limit it in order to avoid environmental collapse. Along these lines was the well-known report to the Club of Rome that we have already been cited (Meadows et al. 1972). As in Ehrlich's book, most of the predictions have proven to be wrong. More scientifically rigorous is the concept of '*Planetary boundaries*' developed by Johan Rockström of the *Stockholm Resilience Center* and Will Steffen of the *Australian National University*, which sets the limit values that different environmental indicators would have for the Earth system to maintain its natural functioning (Rockström et al. 2009; Steffen et al. 2015). Of the nine planetary boundaries defined by these authors, we have already crossed four: climate change, biodiversity, biogeochemical flows and deforestation. The others are drinking water use, chemical pollution, atmospheric aerosols, ozone layer depletion and ocean acidification.

The second aspect of this controversy, and the most ethically sensitive, is what to do with the 'excess' population, i.e., the people who surpass the planet's supposed carrying capacity. One way to adjust them in the long run would be to drastically reduce population growth. Paul Ehrlich, asked for a president for the U.S. 'with guts to say that no one has ever given a reason why there should be more than 140 million Americans alive at any one time', but he did not give strong arguments to defend that figure, nor did he indicate what to do with the 180 million Americans who exceed it. He went on to point out that 'The patriotic thing to do would be to limit the number of children'. In his view, couples should 'stop' at the second child. 'Having more is selfish and irresponsible' he said.

This kind of attitude is found among well-known naturalists, although some have been more explicit in their statements. For example, Jacques Cousteau, the well-known oceanographer, declared in 1991 in an interview with a UNESCO magazine: 'It is terrible to have to say this. The world's population must stabilize and to do that we must eliminate 350,000 people per day', and then went on to say: 'This is so horrible to contemplate that we shouldn't even say it. But the general situation in which we find ourselves is that pitiful' (Cousteau 1991). Cousteau advocated limiting population growth, mainly through policies favoring the education and promotion of women in developing countries. In a similar vein, James Lovelock considered that limiting human growth is an ethical imperative to mitigate the pressure that human beings exert on the planet: 'those who live in harmony with the environment favor the selection of their offspring' (Lovelock 2006). The microbiologist Lynn Margulis expressed it in a more drastic way, even implicitly defending the extinction of human beings: 'In this theater of evolution, all actors are equally necessary; if we can do without one, it is the last one to arrive'.

By what means should this population control be carried out? For some proponents of deep ecology, measures should be coercive, even if peaceful means are used (Devall and Sessions 1985). Included in this group is also Garrett Hardin (1968), who coined the concept of 'The Tragedy of the Commons', referring to the collapse that occurs when personal freedom (and selfishness) is favored over the common good. Hardin argued that appeals to conscience are insufficient to deal with the seriousness of the problem of resource scarcity. He advocated for the imposition of fines or higher taxes on those who had more children.[1] He pointed out:

> The freedom to procreate will lead to ruin for all of us. At this time, in order to avoid drastic decisions, many of us are tempted to rely on conscience and responsible parenthood. This temptation must be resisted, because appealing to consciences acting individually implies betting in the long run on the disappearance of all conscience, and, in the short term, on increasing anxiety (...). The only way to preserve and nourish other and more precious freedoms is to renounce the freedom to procreate, and very quickly.
>
> (Hardin 1968, 1248)

Other authors recommend a moderate population growth on the basis of an ethical responsibility with the environmental conservation of the planet. This would be the position of some Christian thinkers (Guillebaud and Moore 2009), for whom environmental balance requires a voluntary effort to restrict the number of children.

Behind these positions is the assumption that the Earth has already reached its saturation limit, both in terms of the quantity and quality of the resources we have available for the development of humans and other living beings. However, the idea of saturation is not as new as we might think:

> We are a heavy burden on the world and resources are scarcely sufficient; complaints come from all sides and needs are continually increasing, even though nature can no longer support us. We must face the facts and accept that hunger, disease, wars and floods are barriers to an overgrown humanity.

One might think that this text was written by a contemporary author, but this is not the case. It is a paragraph taken from the work *'De Anima'*, written by Tertullian around 200 A.D., when the earth 'supported' a little more than 300 million humans, more or less those living in the U.S today. As Marcos, the author who quotes this text, concludes, the correct question is not 'how many humans' the planet can support, but 'how many humans, with what technology, with what customs and with what social order' (Marcos 2001, 121).

In fact, for another group of authors, we cannot properly speak of a fixed planetary carrying capacity, but rather one that depends on the technologies and ways of life at any given moment. For these authors, the planet's resources are not infinite, but they are somehow unlimited, since the same amount of raw materials can elastically give rise to many more goods since technology offers us alternatives to the materials we use at any given moment. We have already quoted the example of copper given by the Jewish economist Julian Simon (1932–1998), who proposed to define scarcity in economic terms, by equating it with rising prices. In his opinion, if the prices of raw materials decrease, we must conclude that they are more abundant, either because new deposits are discovered, or because less of that material is needed to perform a certain function. In the 1980s, there was a renowned intellectual dispute between Simon and Ehrlich, which even led to a bet between the two on the evolution of the prices of certain raw materials. Simon won the bet on all the materials that Ehrlich himself had previously chosen, as a few years later the prices of those materials did not increase.

In short, although the current level of population and its growth pose important challenges for the conservation and ecological stability of the planet, population reduction does not seem to be an adequate response to the problem and may conceal a selfish attitude of rich countries afraid to share their standard of living with developing societies. Here it seems appropriate to quote a text from Pope Francis' encyclical LS, which provides the vision of this problem in developing societies:

> Blaming the increase in population and not the extreme and selective consumerism of some is a way of not facing the problems. This is intended to legitimize the current distributive model, where a minority believes it has the right to consume in a proportion that would be impossible to generalize.
>
> (LS, n. 50)

We agree with this approach. It is not so much the number of people on the planet, but the way they live. For example, if a resident of the United States consumes ten times more energy than a resident of India, the population of the United States (330 million in 2024) would have twice the environmental impact of India (1.4 billion), even though it is almost five times smaller. In short, the environmental impact is highly dependent on the standard of living. Ghandi's maxim that 'The world has enough for the needs of all, but not enough for the greed of a few' could be inserted here.

On the other hand, we must not forget the enormous suffering caused by coercive population control policies, particularly in China, where the one-child policy was in effect for several decades, including the widespread practice of forced abortions, forced sterilizations, fines and imprisonment. Today, the demographic consequences of this policy (gender imbalance, population aging, sustainability of social services) are very much present in Chinese society, which has modified its one-child policy in recent years to include economic incentives for larger families.

Of course, the demographic situation varies greatly from country to country, but it is striking that the most pro-population control ideologies are being promoted in the wealthier regions with the lowest fertility rates. The promotion of contraceptive methods in industrialized countries with fertility rates below the replacement rate seems to be dangerous for guaranteeing future social services. In addition, those countries more concerned about the environmental impacts of human population seem indifferent about the repercussions of the fast-growing rate of pets' population. In a pioneering study published a few years ago, Gregory Okin (2017) estimated that environmental impact of the food consumed by the 163 million dogs and cats (half of the US human population), living in the US citizens' homes. He calculated that the dietary requirements of those animals accounted for 25±30% of the country's environmental impacts from animal production in terms of the use of land, water, fossil fuel, phosphate, and biocides. To this figure should be added the rest of the requirements of these pets (health care, grooming, transportation, etc.).

Finally, those pro-population control positions almost always ignore the fact that humans not only consume resources, but also generate them, among many other contributions to social welfare. A new child will bring more consumption, but also new ideas that may the critical to create vaccines or solve illness, new energy sources or new building techniques more respectful with the environment. A pet will not.

In summary, when we consider that the solution to environmental problems is to reject all human intervention in nature, we seem to forget that we are also part of nature. Humans are not aliens on Earth, but part of the natural evolutionary process, and therefore we receive and provide exchanges of matter and energy to nature. This alleged alienation from nature is precisely one of the roots of environmental problems. Being aside from nature is not going to solve environmental problems, but rather the opposite: we need to emphasize the sense of belonging to the natural system, being further committed to its conservation and restoration, considering it as our home.

The Political Answer

Politics is the art of governing and answers the question of how to live together in a society. In the Middle Ages, the guiding worldview of governance and human power was limited by the will of God. Thus, all decisions about how to govern the earth and society had to be based on divine justice and wisdom, according to the Aristotelian concept, adopted by Thomas Aquinas, of the subordination of private interest to the common good, in which the two in some sense coincided. Politics was seen as a means at the service of society and each individual. In the Modern Age, the development of the scientific method attempted to decipher the divine code by seeking the principles of nature to dominate it without limit. Technological development enhanced human capacity for transformation, leading to economic prosperity, but also to greater social inequalities and conflicts, as well as new environmental problems.

The outbreak of ecological issues in national politics followed a gradual process, first encouraged by pioneers of environmental concern (see Chapter 3) and then fostered by the political outraging following the social protests of May 68. Until then, political ecology was focused on regulations for the protection of species, natural areas or pollution control, but without a unitary approach to political action. In this breeding ground, the creation of the Club of Rome in 1970 and the holding of the United Nations conference in Stockholm in 1972 marked the beginning of international environmental policies, as well as a demonstration of the influence that the environmental movements were gaining. Nowadays, ecological issues have been introduced into practically all areas of territorial planning, at least in developed countries, where nature conservation forms a basic pillar of what we nebulously refer to as *progress*. In European governments, for example, decision-makers of all political persuasions have been in favor of strengthening environmental protection legislation and have led the world in a greater commitment to the sustainability of the planet, from climate change to forest conservation or improving access to water.

Political ecology has been developed around three different positions, closely related to different political theories. On one hand, liberalism, which assumes that free market will solve the main problems and therefore the state should aim to promote personal freedom and avoid unnecessary regulations; on the other, socialism, which emphasizes the common good and the state intervention to overcome inequalities and guarantee minimum quality standards (for instance, in terms of air or water quality, food production, industrial processes or recycling practices). Finally, a third way, mixes both political approaches, aiming to favor personal freedom while respecting a set of standards coordinated by governments.

Analyzing different environmental policies, Gudynas (2014) distinguishes four stages regarding the consideration of sustainability in national policies:

No consideration: The leading idea is economic growth and ecology is considered as a limitation of development.

Weak: Some consideration of environmental issues, limits are set but not enforced. Emphasis is placed on ecosystem services and natural capital.

Strong: Technical and political approach. The idea of unlimited growth is seriously challenged. Nature is not primarily considered in terms of economic benefits, with a clear goal of preserving natural capital.

Very strong: Substantial criticism of the idea of unlimited growth, new styles of development, nature as patrimony, biocentric ethics, political approach. Ecological citizen is fostered.

Political ecology can be developed at different scales: individual, considering personal impacts; enterprises, considering production and distribution processes, commonly through the idea of corporate social responsibility; community, through NGOs and environmental volunteers; national, including environmental approaches at all levels of government, and international, coordinating efforts with other countries to overcome global problems.

The international scales have been greatly developed in the last decades, particularly after the celebration of the *Earth Summit* of Rio de Janeiro (1992), where relevant UN international treaties were signed (particularly those leading to combat Climate Change, Biodiversity Loss and Desertification). Since the Rio Summit, successive international meetings have led to a convergence of human development and environmental policies, culminating in the 2000 Millennium Development Goals (MDG) for poverty eradication, equity and development, and conservation. Particularly active in terms of international agreements have been the UN Framework for Climate Change (UNFCC), which entails that all governments that signed the agreement meet every year to discuss common climate policies. These Conference of Parties (COP) of the UNFCC have been critical to reach certain consensus on climate change mitigation, although still the goals are not very ambitious. The main treaties were agreed in 1997 in Kyoto (where all developed countries agreed to set GHG emission limits), and in 2015 in Paris (where all countries agreed to develop national plans to reduce emissions, although specific targets were not established). Other environmentally oriented treaties at international level have been the Montreal protocol (banning the production of chemical components that damage the stratospheric ozone layer), and the update of the MDG, in 2015 leading to the currently active Agenda 2030, establishing 17 Sustainable Development Goals (SDGs) with specific indicators, which include not only environmental aspects, but also poverty eradication, health, education and social protection policies.

Another branch of environmental policy has emerged in response to the conflict that is often latent in environmental management and policy decisions. Hence the emergence of a form of policy called *environmental governance*, which focuses on the resolution of environmental conflicts. In addition to state actors, there is also space for and involvement of private political actors, associations, and environmental lobbies. Currently, environmental governance is an expanding form of politics aimed at achieving nature conservation at both the international and national levels. Understanding how environmental conflicts arise, are resolved, or transformed is part of political theory models. Floyd (1993) suggests distinguishing between conflicts arising from misinformation, conflicts arising from the confrontation of interests, and conflicts arising from the confrontation of fundamental principles. In a system of freedoms, dialogue and negotiation are posited as optimal solutions for the first and second cases, since in the first case they lead to a win-win situation through mutual understanding, or where each party partially satisfies its interests through compromise. The problem arises when this political model is applied to a conflict motivated by immutable principles or values rather than interests. Anthropocentric or biocentric alternatives may represent political principles that are difficult to negotiate. A strong anthropocentrism in the political realm assumes that no criteria beyond the popular and usually utilitarian will should be recognized, so that the majority value should be taken as the best achievable outcome in itself. From a biocentric perspective, the political value to be accepted does not always have to coincide with that of the majority but must be judged in terms of what is closest to the structure of the ecosystem or nature according to scientific knowledge.

Integral Ecology

In our opinion, the current environmental crisis will not be solved by the above indicated answers, as none of them is completely satisfactory. What is needed is a more profound change that modifies the way we relate to nature, which entails a change in our current scale of values, giving priority to postmaterialist values over the current utilitarian view that deteriorates the environment and deteriorates ourselves. This change of values has been well expressed by several Christian leaders (mainly the Patriarch Bartolomew and pope Francis), under the term 'ecological conversion', which is equally valid for people of any philosophical or religious orientation, and would have the following characteristics:

It should affect our relationship with nature and with ourselves (human ecology).

It would mean changing our values, avoiding superfluous consumption, which would lead to the promotion of new lifestyles based on the search for truth, beauty and goodness.

It would imply a new relationship with other human beings, based on communion for common growth.

In our opinion, these elements are essential to achieve the harmony that both the natural world and our own physical, emotional and spiritual health need. Let's review some practical implications of this attitude.

Human Ecology: Towards an 'Ecological' Morality

We began this book by talking about the concept of nature, of what we consider sufficiently natural to deserve conservation. If, from a biological point of view, what is natural is the most desirable state (regardless of whether one considers it to be the fruit of a random or of a convergent evolution), then respecting the natural state of things should be a criterion for action, that is, it should have ethical or moral force. Why are certain actions good or bad? What is the 'canon' against which we can compare them in order to make a moral judgment of our actions?

For many centuries, and still in many countries, the most widely followed moral criterion is the one that conforms to the sacred texts of a given religion, which are considered by believers as a living testimony of divine revelation and a reference for ethical action. The most obvious case would be the Ten Commandments, the synthesis of Judeo-Christian morality. Since these commandments are open to different interpretations, each religious tradition recognizes an authority who discerns their meaning for the circumstances of each historical epoch. This is the role of religious teachers (rabbis, ulemas, bishops, theologians, gurus...), who offer a deeper interpretation of the texts. If there is no single magisterium, within the framework of each religion there can be multiple interpretations of the revealed texts (this is the case of Protestant Christianity). For Catholic or Orthodox Christianity, the historical tradition and the constituted religious authorities would mark the key for a certain interpretation of the biblical texts.

In Western culture, the acceptance of this religious authority as a source of morality is currently quite weak. Autonomous judgment has replaced external judgment, and the criteria of morality become a manifestation of the personal values that each person has, which can also change according to circumstances. This approach easily leads to moral relativism, in which anything is acceptable or unacceptable according to the values that each one has, perhaps with the only limitation of 'not offending others'. However, if the ethical criteria for classifying something as good or bad were subjective, social coexistence would be almost impossible, so states legislate normative criteria to ensure peaceful coexistence among citizens. The law, in this view, would simply be an 'agreement of minima', and a certain behavior would be prohibited not because it is objectively bad, but because society has defined that behavior as inappropriate. In other words, this approach – commonly referred to as 'legal positivism' – implies that the criterion of goodness is what the laws say at any given moment. Since, in this view, legality is the result of social agreement, the qualification of goodness from badness would be transient, and something that is illegal at one moment might be approved at another.

Our purpose here is not to enter into a discussion of legal philosophy, but simply to point out the implications of choosing one or another criterion for ethical judgment. Since the one based on sacred texts would not apply to non-believers, and the subjective criterion (one's own values) seems unsatisfactory for social organization, only the legal dispositions would be common to all citizens, which still has the drawback of being historically relative. But there is another alternative for establishing common moral criteria: the nature of things. This criterion has been peacefully accepted in Western history until relatively recent decades and has been called 'natural law'. The growing concern for the preservation of nature may lead to a revaluation of this concept, since – if we want to preserve what is natural – it seems reasonable that naturalness should be considered something good, a basis for moral judgment. Several intellectuals have made similar proposals to overcome the moral relativism that seems to pervade Western society. Benedict XVI, for example, in his address to the German Bundestag, pointed to the growing importance of ecology as an objective pillar of ethical judgment:

> ... the appearance of the environmental movement in German politics from the 1970s onwards, although it may not have opened the windows, was and is nevertheless a cry for fresh air, a cry that cannot be ignored or rejected because too much irrationality is perceived in it. Young people realized that in our relations with nature there was something wrong; that matter is not only a material for our use, but that the earth has in itself its dignity and we must follow its indications.
>
> (Benedict XVI 2011)

He emphasized that human action cannot be a pure indetermination of freedom, but, like nature as a whole, is subject to indications inscribed in its very nature. This also affects moral action, because:

> Man also possesses a nature that he must respect and that he cannot manipulate at will. Man is not merely self-creating freedom. Man does not create himself. He is intellect and will, but he is also nature, and his will is rightly ordered if he respects his nature, listens to it and accepts himself for who he is, as one who did not create himself. In this way, and in no other, is true human freedom fulfilled.
>
> (Benedict XVI 2011)

Finding objective moral foundations on the nature of things would be applicable to everyone, regardless of his philosophical and religious values, and therefore would be a sound basis for social coexistence, beyond conjunctural agreements (International Theological Commission 2009).

Here the parallelism between moral action and the philosophical and biological foundation is clear: if we want to be more respectful of the natural environment, we must understand how nature works, since deviations from the natural state can have negative consequences on the system as a whole. Therefore, the relationship between ecoethics and bioethics should be much closer, as several authors have already pointed out (Gómez-Heras 2012; Turvey 2016).

There are many examples of arbitrary transformations of nature which have led to undesirable effects that alter the original equilibrium and implied in same cases severe damages. Over-fertilization or wide application of pesticides in agriculture have entailed pollution of soils and water; unnatural feeding of livestock to enhance rapid growth conducted to the *mad-cow* crisis. The same applies to the use of artificial preservatives or other substances that improve the shelf life or presentation of food but may end up causing health risks (some of them are even carcinogenic). For this reason, the consumers are now growingly appreciating naturally cultivated or grown food.

The same principle could also be applied to moral issues that affect the way we human beings live, and where it is important to recover a deep human ecology that leads us to qualify as good or bad what is or is not proper to our nature. If the use of certain preservatives, herbicides or pesticides is artificial – and therefore potentially harmful to human beings – why not also, for example, the use of contraceptives? Are these more natural than the former? Should the concern, so often reasonable, about the impact of transgenic crops on human health be less obvious than the application of cloning techniques or genetic selection in humans? If we force an herbivore animal to feed on meat, we are altering its natural state and will have negative side consequences, as evidenced in the *mad cow* crisis. Similarly, if we introduce technology that alters effects that are a direct consequence of human nature, the effects can also be catastrophic.

In December 2011, we organized an international colloquium on the dialogue between Science and Religion. One of the speakers, an English neonatologist, showed the extent to which artificial intervention in the beginning of human life has led to far-reaching ethical consequences (Wyatt 2012). He gave the example of a single American woman who wanted to become a mother, although she did not have a partner and did not want to get pregnant. She searched the internet for a sperm donor, an egg donor and a surrogate mother. In short, she ordered a child who had nothing of herself except the money she had invested in the orders. This is obviously an extreme case, but it serves to illustrate how reproductive techniques have changed the nature of motherhood. Instead of accepting a baby as a gift, the fruit of a natural relationship between a man and a woman, the human being is now able to exercise a dominion over reproduction that turns the baby into a commodity, an object that can be commercially commissioned, as if that child had no right to an autonomous life project.

It seems to us that respect for 'human ecology' is an indispensable part of our commitment to the environment, if we are consistent with its deepest philosophical or theological roots. In short, respect for nature begins with respect for our own nature, by accepting the natural course on which our life project unfolds. Human beings, of course, have the capacity to make decisions about their future, but we are not indeterminate beings, we do have limitations. Certainly, many of us would love to fly, but we cannot do it, because it is not in our nature, we do not have wings: we can do it with an airplane, but it is not we who fly, but the airplane. We could build mechanical wings and facilitate that task, but it would still be something external to our nature, whereas for a stork or an eagle it is perfectly natural. Such artificial wings would require maintenance, would cost extra effort to use, and would be in danger of falling off and bringing us down. Birds do not need anything artificial to fly, they do it naturally and that is why they do it very well, and there is no negative consequence to doing so. Following this example, it is obvious that, if something is natural to our physical and spiritual constitution, it will be much easier to approach it and it will also be much more beneficial to our physical and mental health.

It is not a matter of not using medicine and letting things resolve themselves by natural means, but rather of intervening only to solve problems that involve a disruption of natural functioning. For example, we may operate on a cow that is unable to ruminate in order to solve this problem, simply because the nature of 'being a cow' implies ruminating (which they do splendidly, by the way). Similarly, we find it morally acceptable for medicine to provide artificial technical means to cure a disease, because any disease is basically a malfunction of a natural process. Eyes, for example, serve to see, and they are magnificently designed for this task (few artificial cameras have such optical resolution). If they do not perform that function well, we proceed to a surgery or choose to wear glasses, as we are trying to recover the eyes' natural purpose. It is not the same case, morally speaking, when we use medicine to prevent someone from being fertile when they are naturally fertile (sterilization) or from dying forcibly (euthanasia). It is worth noting that contraceptive surgeries (tubal ligation or vasectomy, for example) are the only ones performed by medicine to stop something that naturally works well. We understand that artificial intervention in a process that is natural would only be ecologically admissible when it attempts to correct a defect in natural behavior (for example, an extrauterine pregnancy), not when it avoids a natural consequence (normal pregnancy), whether desired or not, since the desire of the parents is not morally stronger than the right to life of the conceived child.

Of course, this implies overcoming the utilitarian mentality, where ethics is equated with efficacy. Artificial manipulation of human embryos, for example, may be medically beneficial for someone, but that is not the main moral criterion for accepting it or not. What is important is not so much whether they will be useful or not, but whether they will ultimately benefit the human species. Sometimes immediate benefits hide big problems in the medium or long term. Removing forest cover from a steep watershed may provide immediate agricultural yields, but what will happen when there are torrential rains? Intense soil erosion is likely to occur, which can lead to landslides and flooding downstream. Initial agricultural improvements would end up being very short-lived.

From an environmental perspective, this would be as objectionable as the artificial selection of crops or the installation of a nuclear power plant, which can also have positive consequences from a pragmatic point of view. Why is efficiency or economic benefit approved in some cases and not in others? Is it natural to use human beings, without their consent, to cure others? Is it ecologically justified to convert a human embryo into a medicine for another person? Again, the ecological postulate that the artificial can turn against our nature applies, since many times embryonic stem cell treatments have often resulted in rejections or carcinomas and for this reason there are currently very few viable treatments from these cells (Wuputra et al. 2020). Adult stem cells, on the other hand, have led to many more successful therapies, have no rejection problems (they are often taken from the same patient) and do not involve the destruction of human embryos. Same can be said about natural versus artificial methods of family planning, the former completely exempt of side effects (Genazzani et al. 2023).

If we take the natural argument to its ultimate consequences, we will tend to reject artificial interventions on nature -human or biological-, as long as they do not attempt to correct some natural dysfunction. In short, whether something is natural or not can be considered a criterion for moral action, which not only concerns our relations with the environment, but also respect for our own body, their biological and affective functioning.

Occasionally, critics of environmentalist positions use as an argument that certain ecological ideologies defend the protection of animal species while favoring abortion and euthanasia. These ecological ideologies may not be aware of the connection between ecoethics and bioethics, and may be inconsistent with respect to the protection of life. Of course, there are many other people with both environmental and bioethical concerns who defend the life of the unborn or the terminally ill. To present the defense of nature and human life as antagonistic does not help either ideal. What matters is whether or not they are 'par essence' contradictory. In our view, they are complementary rather than incompatible. Those who love all life should also love human life, either already conceived or at the end of its natural process. Likewise, respect for the gift of life concerns all creatures, and a truly pro-life stance that favors human life should also support the life of other creatures. Artificial abortion is unique to the human species (no other species causes the death of already conceived offspring). Among other arguments, the ecological reason for proposing a true culture of life is undoubtedly noteworthy.

In fact, it is completely unnatural to attack a human being in pregnancy in the most ecologically appropriate place for it: the mother's womb, which is the primary sanctuary of life.

Changing Social Structures

Several authors have shown the link between nature conservation and peace. One of the first moments in which nature conservation was introduced in the political agenda was the President of the United States, Theodore Roosevelt, since his childhood a friend of nature and an admirer of the conservation movement, which had its origins in Thoreau and Muir. Initially, this environmental political concern was focused on environmental health issues and was later strengthened by Rachel Carson's denunciation in her book 'The Silent Spring' in early 1960s. Ten years later, in 1973, Ernst F. Schumacher pointed out that a more sober way of life, one that pays more attention to spiritual and cultural goods and less to material ones than to the mere accumulation of resources, not only does not bring happiness in the long run, but is a certain obstacle to achieving it. In addition, Schumacher pointed out the effects that a globalized economy, which only takes into account the interests of the economic powers, would have on local economies, creating a social and cultural divide between those who participate in these global trends and those who do not. Walking through the urban center of an Andean country, one can observe examples of the dual economy that Schumacher speaks of, with young people in the same avenue who are completely adapted to a 'Western' way of life, in the way they dress or in the technologies they use, and others of the same age who maintain the way of life and dress of their traditional cultures: in the same space, two ways of life are so far apart that they seem to be two different worlds, what he call 'dual economy':

> It is not just a matter of some people being rich and others poor, even though they share a similar way of life; it is a matter of two ways of life so diverse, coexisting in the same place, that even the humblest member of one has a daily income that is a high multiple of that available to the hardest worker in the other. The social and political tensions arising from the dual economy are so obvious that they require no detailed description.
>
> (Schumacher 1973, 175)

These dual economies are the result of an economic model that prioritizes the profit of the few over the happiness of the many, degrading human beings and the environment in which they live. They are also the ultimate cause of instability and conflict. Schumacher affirms that peace should be built upon meeting basic needs with sobriety for oneself and others, generating more stable societies:

> The cultivation and expansion of needs is the antithesis of wisdom. It is also the antithesis of freedom and peace. Any increase in needs tends to increase one's dependence on external forces over which we have no control, and consequently increases existential fear. Only by reducing needs can we promote a genuine reduction of tensions, which are the ultimate causes of violence and war.
>
> (Schumacher 1973, 34)

In short, environmental awareness also means underlining the link that binds us to nature and to other human beings. Social and natural degradation are part of the same process and should be overcome with the same weapons. Building world peace would not only require mitigating existing economic and social inequalities, but also reducing the illicit exploitation of natural resources that arise at all levels, local, national, transboundary or global. The main international organizations now recognize that in order to prevent conflicts it is necessary to have a fair distribution of goods that allows the stabilization of peoples, as well as an awareness that natural resources do not know borders, recognizing that social stability is built by granting a dignified life to people, which includes a healthy and integral environment. The conflicts in many developing countries, full of natural resources but mired in poverty and dependence on the old and new metropolises, bear witness to this.

Extreme poverty goes hand in hand with environmental degradation, and often with disregard for human beings, which generates social tensions and creates vicious circles that can only be broken if solutions are found in all spheres of the problem. For this reason, solidarity becomes a tool not only for justice but also a key element for the conservation of nature. Social and environmental problems go hand in hand and peace cannot be achieved without the protection of human life and nature.

The great religions have linked human dignity, the preservation of nature and peace. The social values of Christianity, for instance, provide guiding principles of subsidiarity, solidarity, the common good and the universal destination of goods. It is no coincidence that several of the first messages of the Popes on the environmental issue were produced precisely on the World Day of Peace, which the Church celebrates every year. In his 1990s message which John Paul II entitled 'Peace with God the Creator, Peace with all of Creation', he stated:

> Today there is a growing conviction that, in addition to the arms race, world peace is threatened by regional conflicts and the injustices that still exist among peoples and nations, as well as by the lack of due respect for nature, the unregulated exploitation of its resources and the progressive deterioration of the quality of life. This situation provokes a sense of instability and insecurity that in turn favors forms of collective selfishness, hoarding and prevarication.

> Faced with widespread environmental deterioration, humanity realizes that it can no longer use the earth's resources as in the past. Public opinion and political leaders are concerned about it, and scholars of the most varied disciplines are examining its causes. An ecological conscience is thus being formed, which should not be hindered, but rather encouraged, so that it may develop and mature, finding adequate expression in concrete programs and initiatives.
>
> (John Paul II 1990, n.1)

In the same line, Patriarch Bartholomew I linked the respect of nature (Creation) and respect to human rights, as two intimate union realities:

> We cannot separate our concern for human dignity, human rights or social justice from concern for ecological protection, preservation and sustainability. These concerns are forged together, comprising an intertwining spiral that can either descend or ascend (...) The most serious contemporary threat against such a culture of solidarity is the prevailing economy – what we might call, the fundamentalism of market and profit. We are not qualified economists, but we are convinced that the purpose of economy should be for the service of humankind. It is not by coincidence that the terms economy and ecology share the same etymological root.
>
> (Bartholomew 2015)

Care for nature also involves care for people, the ability to reflect on the impact of our behavior on others, on the natural and human environment, to look beyond individualistic position, to reinforce the vision of common good. It is not a matter of moving towards central-government positions, increasing the regulations, but rather to strengthen the political and social structures that pursue the common well-being of all citizens and all nations, overcoming the trends of those who still consider the free market the panacea of all problems. Freedom should not be used to justify selfishness, nor common-good to back state mentalities that aim to regulate all aspects of social life. Subsidiarity and solidarity, common good and human dignity, should be complementary, not contradictory.

Changing Lifestyles

The lifestyles we adopt are key to explaining our environmental impacts, since all things we consume need a certain quantity of resources and energy. A person biking or using public transport to get around uses a fraction of the energy consumed by someone in a large car doing the same distance. Similarly, the food we eat has a big impact on our resource consumption, as different diets require very different amounts of water and energy. The same can be said for every other area of our consumption habits, from clothing to computer equipment, from tourism to climatization.

The concept of environmental footprint has received a lot of attention in the media as a way to concretize citizen's commitment to the environment (Gaussin et al. 2013). It involves considering two concepts: on the one hand, that of Life Cycle Assessment (LFA) and on the other, that of the type of footprint itself. It is worth remembering that environmental impacts should be accounted for during the whole life of the product we are measuring: from the beginning to the end of the processes require to obtain that product. LFA analysis includes a collection of methods that allow measuring the overall footprint of a certain activity or product. An important distinction in this regard indicates to extent or scope to which this estimation is made, that is over which stages the calculation is to be made. The most common scopes in LFA are the so-called *cradle-to-gate or business-to-business (B2B)*, and the *cradle-to-grave or business-to-consumer (B2C)*. The former (B2B) includes all phases from the production, transport and storage of the product until it is purchased by the consumer or by another company; while B2C the scope adds to those stages the processes linked to the use and possible recycling of that product (Sala and Castellani 2019). (See Figure 7.1)

The second dimension of environmental footprint indicates the type of resource that is being measured. The most common are the water footprint and the carbon footprint, which respectively indicate the water and emissions of GHG required to obtain a given product or service. The former is measured in liters of water and the second one in kg of CO_2 equivalent. The ecological footprint concept has also been widely used, indicating the surface

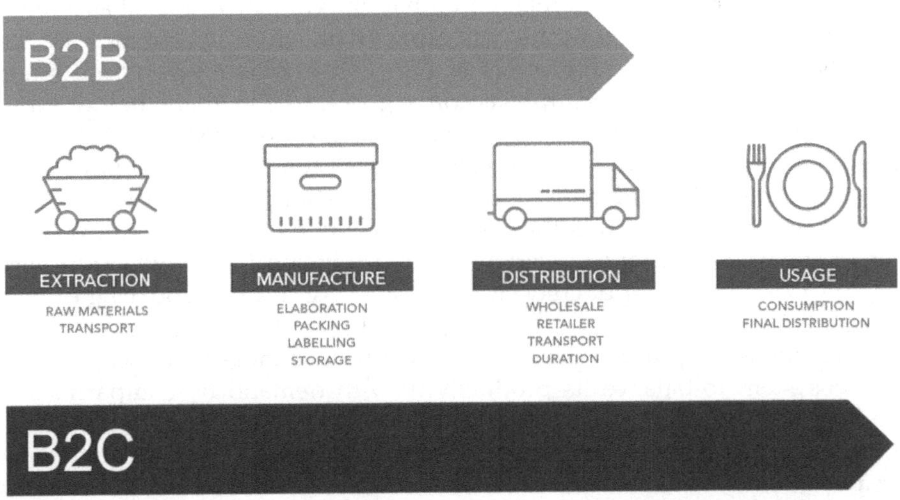

FIGURE 7.1
Graphical representation of the scopes of the Life Cycle Assessment for environmental footprints. (https://huellaco2.org/en/ciclo-vida.php, reproduced with permission.)

area that would be required to produce the resources consumed annually by a given person, region or country (Collins et al. 2018).

The environmental footprint of the same product can vary greatly depending on how it is produced. For food, this means how it was cultivated, harvested, transported, processed, stored and distributed. In the case of the water footprint, for example, the amount of water needed for the full LFA of an average kilo of bread is 1,608 liters, that of tomatoes is 214 liters, while 4,325 liters are required to produce 1 kg of chicken meat or more than 15,400 liters for the same weight of beef (https://waterfootprint.org/resources/interactive-tools/product-gallery/ last consulted in August 2024). However, it is worth noting that these values are averages, since the calculations refer to very different forms of production. Ideally the footprint should be calculated for each product and from the specific processes that were used to obtained. The amount of water required for producing beef is not the same when the beef comes from extensive farms, where the animals eat from natural grass, that from an intense one which requires food imported from other areas.

One of the most developed environmental impact assessments of personal consumption has been the calculation of the carbon footprint (CF), as is closely linked to climate change mitigation. The CF measures the amount of greenhouse gas (GHG) emissions generated during the life cycle of a given product or activity. As indicated earlier, the most prominent GHGs are CO_2, CH_4 and N_2O. Since emissions can be from different GHGs, in order to have a common scale of emissions, CF is usually expressed in terms of mass (grams, kilograms, tons...) of CO_2 equivalent (CO_2e). This concept quantifies total emissions based on the Global Warming Potential (GWP) (also called radiative forcing) of each GHG in relation to that of CO_2, which is used as a reference. In short, the GWP indicates the capacity of a given gas to increase the greenhouse effect compared to that of CO_2, considering a given time frame (usually 100 years). For example, according to IPCC, methane has 28 times more GWP than CO_2, while N_2O has 265 times more GWP (Myhre et al. 2013). The average values of CF are calculated on scientific studies that have done life cycle analysis in different countries (Burgui-Burgui and Chuvieco 2020; Ivanova et al. 2017). As was the case of water footprint, the CF of products varies greatly. For instance, as an average, producing 1 kg of wheat implies 0.5 kg CO_2e, while 1 kg of chicken implies: 3.9 kg of CO_2e and 1 kg of beef 27.1 kg.[2]

As for the ecological footprint, in a way it tries to measure the capacity of the ecosystem to balance its productivity with demand by comparing the land area of the country with the land area it needs to produce the goods it uses. If the area required for the consumption of a given country is much greater than its actual surface area, one can say that this country maintains a way of life that is out of balance with its environment. Certainly, world trade makes it possible to obtain goods from other places, balancing one's own resources with others that we can call intangible (innovation, culture, aesthetics...), and which are considered more valuable than the resources themselves that they use, which explains why many advanced countries use more resources than their territory produces. However, for a global whole, this logic would not be viable, since the carrying capacity of the earth system has physical limits, which can be expanded with new technologies, but which are not infinite. The ecological footprint is not only related to a country's population, but also, and mainly, to its standard of living and consumption habits: a nation with a high income tends to consume many more resources than one with a low income. The United States, for example, has a per capita ecological footprint seven times higher than that of the poorest African countries.[3] However, there are countries that have living and consumption habits that allow them to reconcile high human development with a less aggressive impact on the environment. For example, Germany has an ecological footprint value of about 4.7 hectares per inhabitant, much lower than that of the USA (about 7.5 hectares per inhabitant), even though it has a higher Human Development Index.

It is clear that our habits and our consumption greatly influence the amount of resources we use, and it is also evident that we have the capacity to significantly reduce this expenditure while maintaining a reasonable standard of living. There are multiple daily activities, choices we make every day, that have great importance in the long run on our environmental impact: from the means of transportation to the frequency with which we change our household appliances, to the food we consume or the energy we use to maintain a comfortable temperature level. Some data can help us to become more aware of the resources we use every day:

A small car consumes about 4.5 to 5.5 liters of fuel per 100 km; a large car up to 9–12 liters for the same distance. In short, with a small car one can travel almost twice as far as with a large car for the same energy consumption. Generally, the older the car, the less efficient it is in terms of energy consumption.

The CO_2 emissions of a normal car are around 180 g/km (they can reach 400 g/km for large cars). A hybrid car can reduce them to less than 100 g/km. Therefore, for the same distance, the emissions of a normal vehicle are almost double and those of a large displacement vehicle are four times higher than those of a hybrid, with the same level of comfort. From this point of view, an electric car is a much better option (direct emissions: 0 g/km; indirect, depending on the source of electricity). According to the LFA of the car, a mid-size electric car charged with renewable energy will save up to 75% of the emissions of a gasoline car of the same size over the entire life of the vehicle (https://huellaco2.org, last access August 2024).

Our choice of meals impacts our consumption of water and emissions: a higher consumption of vegetables compared to meat products implies a drastic reduction of the water and carbon footprint, since, for example, the amount of water needed to consume 1 kg of beef is 15 times higher than that needed for the same weight of fruit or vegetables. This is not to say that we all need to be vegan, but certainly reducing calf meat consumption (especially from farmed animals) will help us reduce the environmental footprint of food, as well as improve our health (Clark et al. 2019).

Using energy-efficient appliances means significant savings on electricity bills and a reduction in the energy we use on a daily basis. An incandescent light bulb (now taken off the market in Europe) consumed between 60 and 100 watts for a level of luminosity equivalent to a 3 watt bulb based on LED technology, which lasts almost 50 times longer in average hours of use. In short, we are saving between 20 and 30 times the energy consumption of lighting simply by changing the bulbs. The same applies to the most commonly used household appliances. A class A+++ refrigerator consumes an average of 150 kWh per year, while a class C refrigerator of similar power consumes about 600 kWh over the same period. In short, it is four times more expensive in terms of energy consumption, together with what it implies in terms of emissions depending on the source of electricity.

For medium-length travels (300–800 km), using a high-speed train implies 4 times less GHG emissions than travelling the same distance by plane, besides being much more comfortable for the passenger, leaving us in the center of the city and having a similar price (depending on the hours and companies, of course).

We could multiply examples of this type that help us to realize the impact of our daily consumption choices. Certainly, an attitude of responsible consumption will help to reduce our ecological impact, with few changes in our way of life, simply by taking the alternatives with the least environmental impact. A further step would also be to modify our lifestyle, avoiding unnecessary consumption or travel, simplifying our needs, which are often more subjective (apparent) than real. In addition to the environmental benefits that this attitude will bring with it, it will also benefit our spiritual health, which is so often dulled by a senseless materialistic consumerism.

In this area, philosophical and religious motivations may help to reduce consumption. Living frugality, consuming what is really necessary, walking or biking more, will benefit both the planet and our health. Encouraging a simple life has always been present in environmental leaders. For instance, Throreau indicated that

> Most of the luxuries and many of the comforts of life are not only not indispensable, but positive obstacles to the elevation of mankind...the more of these things you have, the poorer you are.
>
> (Thoreau 1854, 8)

In a similar way recent religious leaders remind us that what makes us happy is not materialist consumption, but rather personal values, family and friendship, as pope Francis recalls:

> The emptier a person's heart is, the more he or she needs things to buy, own and consume. It becomes almost impossible to accept the limits imposed by reality.
>
> (LS, n. 204)

Material goods should serve to satisfy our needs, not to create new ones, with habits that make us nobler, more generous, that help us progress as human beings. Moreover, this attitude towards material things will have a direct impact on the ecological footprint of our activity, directly marked by our consumption habits. Therefore, although technological innovations may help to alleviate some of the environmental problems we are currently facing, the most profound solution explicitly refers to changing our ways of living, overcoming a materialistic consumption that grounds happiness in having rather than being.

Exercise for Reflection

There are currently many tools to measure the carbon footprint (CF) of daily consumption, which has been observed as a convenient tool to make lay citizens to be more aware of the climatic impacts of their common habits. We recommend using a carbon footprint observatory that we developed a few years ago (https://huellaco2.org/en/index.php). The site includes a CF calculator, but also basic concepts on what is climate change, CF, LFA and is documented with multiple references. One of the most interesting features of this site is the compilation of CF values for a wide variety of products and activities related to common behavior of ordinary citizens. These tables were obtained from a wide range of sources, including peer-reviewed papers, technical reports and official CF assessments of companies. These tables can be used to simulate the impact of different activities before doing that, and therefore establish the potential CF of various alternatives.

From these tables, several estimations of emission impact can be developed for many ordinary activities. For instance, the teacher may ask the students in small groups (2–3 persons) to organize different alternatives for a 1-week holiday, including different combinations of transportation methods (airplane, train, various types of cars or buses), accommodation places and recreational activities. Then, they may report on the pros and cons of each alternative in terms of emissions, cost and time, and make a decision on which ones are recommended.

Another exercise with the same tool may be related to the dietary selection. Different types of diet may be estimated, including more or less consumption of meat (and with diverse types of meat), fruits or vegetables. The exercise may include all meals during a short period of time (i.e. a week). In addition to the emission estimations derived from the selected diet, the students can compute the calories included in each diet from other sources, and a rough estimation of the costs involved in the different alternatives. Finally, each group may reflect on what it considers the most adequate diet by considering the different aspects involved in food selection.

Notes

1 This policy was already tested in China between the 1980s and 2015, with the one-child policy, as we discuss later.

2 Data taken from the CO2 observatory of the University of Alcalá: https://huellaco2.org/index.php. Last access in August 2024.

3 Data taken from https://footprintnetwork.org/, and corresponds to 2022 information. Last access August, 2024.

8

Final Thoughts

Throughout this work we have seen the roots of the environmental problem, the reasons of all kinds that justify taking this issue much more seriously, and the different responses to this challenge. We have seen that there are many ways to approach the man-nature relationship. An effective personal and social response can be approached from these angles. Now is not the time to continue speculating, but rather to take effective measures, because the environmental crisis requires a forceful response on a scale comparable to the foreseeable impact of the problems.

In a society that tends to compartmentalize reality, that eagerly seeks the responsibility of others to silence a dormant conscience, it is necessary to put an end to dualisms and to seek unity: some distinctions are not only not essential, but are actually fallacious: either man or animal, natural or artificial, preserving a space or developing it. These dichotomies not only prevent us from understanding reality, but also distance us from the personal response that the environmental question demands. The question is not whether we are part of the problem, but whether we are part of the solution.

Perhaps the first step in making operational decisions is to recover our capacity for admiration, for wonder at the beauty of created things, for the fascination with nature that has moved sages of all cultures and times. Environmental ethics finds its inspiring principle in the integrity of creation. We must approach the environment with a sense of belonging, not ownership; with a free mentality, because life, all life, is a gift that we do not deserve. From amazement will come empathy, feeling with others, with other living beings, with other human beings. From the realization that we are part of creation will come the understanding that we are exceptional creatures in this world, without forgetting that we are part of something greater that we did not design, but received as a gift. The result will be to act with greater responsibility and gratitude toward a marvelous nature that is offered to us for our admiration and care.

DOI: 10.1201/9781003545972-8

References

Abbey, E. (2004). *The Monkey Wrench Gang*. London: Penguin Classics. Original from 1975.

Afrasiabi, K.L. (2003). Toward and Islamic Ecotheology. In R. Folz, F. Denny & A. Baharuddin (Eds.), *Islam and Ecology* (pp. 281–296). Cambridge, MA: Harvard University Press.

Agrawal, S. (2017). Geographic Aspects of Vegetarianism: Vegetarians in India. In F. Mariotti (Ed.), *Vegetarian and Plant-Based Diets in Health and Disease Prevention* (pp. 93–106). London: Academic Press.

Armitage, K.C. (2009). *The Nature Study Movement: The Forgotten Popularizer of America's Conservation Ethic*. Lawrence: University Press of Kansas.

Attfield, R. (2018). *Environmental Ethics: A Very Short Introduction*. Oxford: Oxford University Press.

Augustine, S. (413–426). *The City of God*. London: Penguin Classic.

Ballesteros, J. (1995). *Ecologismo Personalista. Cuidar la naturaleza, cuidar al hombre*. Madrid: Tecnos.

Balmford, A., Crane, P., Dobson, A., Green, R.E., & Mace, G.M. (2005). The 2010 Challenge: Data Availability, Information Needs and Extraterrestrial Insights. *Philosophical Transactions of the Royal Society B-Biological Sciences, 360*, 221–228.

Barnosky, A.D., Koch, P.L., Feranec, R.S., Wing, S.L., & Shabel, A.B. (2004). Assessing the Causes of Late Pleistocene Extinctions on the Continents. *Science, 306*, 70–75.

Bartholomew, P. (2015). *Creation Care and Ecological Justice, Address to the Oxford Union*. Available at: www.orth-transfiguration.org/wp-content/uploads/2016/09/HAH-Creation-Care-and-Ecological-Justice.pdf.

Basil, Saint (361–364). *El Hexameron*. Edimburg: T&T Clark.

Baxter, W.F. (1974). *People or Penguins: The Case for Optimal Pollution*. New York: Columbia University Press.

Benedict XVI (2007). *Message for the Celebration of the World Day of Peace*. Vatican: www.vatican.va/content/benedict-xvi/en/messages/peace/documents/hf_ben-xvi_mes_20061208_xl-world-day-peace.html.

Benedict XVI (2008). *Message for the Celebration of the World Day of Peace*. Vatican: www.vatican.va/content/benedict-xvi/en/messages/peace/documents/hf_ben-xvi_mes_20071208_xli-world-day-peace.html.

Benedict XVI (2009). *Caritas in Veritate*. Vatican: www.vatican.va/content/benedict-xvi/en/encyclicals/documents/hf_ben-xvi_enc_20090629_caritas-in-veritate.html.

Benedict XVI (2010). *World Day of Peace*. Vaticano: www.vatican.va/content/benedict-xvi/en/messages/peace/documents/hf_ben-xvi_mes_20091208_xliii-world-day-peace.html.

Benedict XVI (2011). *Address to the Bundestag*. Berlin: www.vatican.va/content/benedict-xvi/en/speeches/2011/september/documents/hf_ben-xvi_spe_20110922_reichstag-berlin.html.

Berry, T. (1988). *The Dream of the Earth*. San Francisco: Sierra Club.

Binde, P. (2001). Nature in Roman Catholic Tradition. *Anthropological Quarterly, 74*, 15–27.

Boff, L. (2012). *El cuidado necesario*. Madrid: Trotta.
Bowman, D.M.J.S., Balch, J.K., Artaxo, P., Bond, W.J., Carlson, J.M., Cochrane, M.A., D'Antonio, C.M., DeFries, R.S., Doyle, J.C., Harrison, S.P., Johnston, F.H., Keeley, J.E., Krawchuk, M.A., Kull, C.A., Marston, J.B., Moritz, M.A., Prentice, I.C., Roos, C., Scott, A., Swetnam, T., Van der Werf, G., & Pyne, S.J. (2009). Fire in the Earth System. *Science, 324*, 481–484.
Brooke, J.H. (1991). *Science and Religion: Some Historical Perspectives*. Cambridge: Cambridge University Press.
Brundtland, G.H. (Ed.) (1987). *Report of the World Commission on Environment and Development: Our Common Future*. New York: Naciones Unidas. www.un-documents.net/our-common-future.pdf.
Burgui-Burgui, M., & Chuvieco, E. (2020). Beyond Carbon Footprint Calculators. New Approaches for Linking Consumer Behaviour and Climate Action. *Sustainability, 12*, 6529.
Cain, M.L., Bowman, W.D., & Hacker, S.D. (2014). *Ecology*. Sunderland, MA: Sinauer Associates.
Calicott, J.B. (1987). *Companion to a Sand County Almanac: Interpretive and Critical Essays*. Madison: University of Wisconsin Press.
Calicott, J.B. (1999). *Beyond the Land Ethic: More Essays in Environmental Philosophy*. Albany: State University of New York Press.
Callicott, J.B. (1995). Intrinsic Value in Nature: A Metaethical Analysis. *Electronic Journal of Analytic Philosophy, 3*, 1–8.
Carr, D.M. (2020). *The Formation of Genesis 1-11: Biblical and other Precursors*. New York: Oxford University Press.
Carson, R. (2002). *Silent Spring*. Boston: Mariner Book. Original from 1962.
Carson, R. (1998). *The Sense of Wonder*. New York: Harper. Original from 1965.
Chargaff, E. (1976). On the Dangers of Genetic Meddling. *Science, 192*, 938–940.
Chuvieco, E. (2012). Religious Approaches to Water Management and Environmental Conservation. *Water Policy, 14*, 9–20.
Chuvieco, E. (2024). *Matter Matters? Extending Christian Materialism and Theology of Work of St Josemaría Escrivá to Ecotheology*. DPhil dissertation: Faculty of Theology and Religion, University of Oxford.
Chuvieco, E., Burgui, M., & Gallego, I. (2016). Impact of Religious Beliefs on Environmental Indicators. Is Christianity More Aggressive Than Other Religions? *Worldviews, 20*, 251–271.
Chuvieco, E., Sánchez-Sorondo, M., & Settele, J. (2016). Religion and Science: Boost Sustainability. *Nature, 538*, 459–459.
Cicero, M.T. (52 BC). *De Legibus*.
Clark, C. (1977). *El aumento de la población*. Madrid: EMESA, Ensayos Aldaba.
Clark, M.A., Springmann, M., Hill, J., & Tilman, D. (2019). Multiple Health and Environmental Impacts of Foods. *Proceedings of the National Academy of Sciences, 116*, 23357–23362.
Clayton, S., & Myers, G. (2015). *Conservation Psychology: Understanding and Promoting Human Care for Nature*. Chichester: Wiley.
Collins, A., Galli, A., Patrizi, N., & Pulselli, F.M. (2018). Learning and Teaching Sustainability: The Contribution of Ecological Footprint Calculators. *Journal of Cleaner Production, 174*, 1000–1010.

Collins, F.S. (2006). *The Language of God: A Scientist Presents Evidence for Belief.* New York: Free Press.

Conradie, E.M. (2006). *Christianity and Ecological Theology: Resources for Further Research*. Stellenbosch: SUN MeDIA.

Costanza, R., D'Arge, R., de Groot, R., Farber, S., Grasso, M., Hannon, B., Limburg, K., Naeem, S., O'Neill, R.V., Paruelo, J., Raskin, R.G., Sutton, P., & van den Belt, M. (1997). The Value of the World's Ecosystem Services and Natural Capital. *Nature, 387*, 253–260.

Cousteau, J. (1991). A Global Commitment (p.13). Paris: The Unesco courier.

Crutzen, P.J. (2002). Geology of Mankind. *Nature, 415*, 23.

Dannenberg, A., & Zitzelsberger, S. (2019). Climate Experts' Views on Geoengineering Depend on Their Beliefs about Climate Change Impacts. *Nature Climate Change, 9*, 769–775.

Davidsen, C., & Kiff, L. (2013). Global Carbon-and-Conservation Models, Global Eco-States? Ecuador's Yasuní-ITT Initiative and Governance Implications. *Journal of International and Global Studies, 4*, 1.

Davis, M.H. (1983). Measuring Individual Differences in Empathy: Evidence for a Multidimensional Approach. *Journal of Personality and Social Psychology, 44*, 113.

Delio, I., Warner, K.D., & Wood, P. (2008). *Care for Creation: A Franciscan Spirituality of the Earth*. Cincinnati, OH: Franciscan Media.

Derrick, C. (1972). *The Delicate Creation. Towards a Theology of the Environment.* London: Tom Stacey.

Deva Diwivedi, K. (2003). Ecological Thoughts in the Vedas. In R. Narayan & J. Kumar (Eds.), *Ecology and Religion* (pp. 3–24). New Delhi: Deep&Deep Publications.

Devall, B., & Sessions, G. (1985). *Deep Ecology*. Salt Lake City, UT: Peregrine Smith.

Diamond, J. (2013). *The World Until Yesterday: What Can We Learn from Traditional Societies?* New York: Penguin.

Dien, M.I. (2003). Islam and Environment: Theory and Practice. In R. Folz, F. Denny & A. Baharuddin (Eds.), *Islam and Ecology* (pp. 107–120). Cambridge, MA: Harvard University Press.

Dobson, A. (2016). *Environmental Politics. A Very Short Introduction*. Oxford: Oxford University Press.

Elliot, R. (2006). Faking Nature. In Light, A. and Rolston, H. *Environmental Ethics. An Anthology*. Malden, MA: Blackwell Publishing.

FAO, F., UNICEF, OMS, WFP (2023). *El estado de la seguridad alimentaria y la nutrición en el mundo 2023*. Roma: FAO.

Floyd, D. (1993). Managing Rangeland Resources Conflicts. *Rangelands, 15* (1), 27–30.

Francis, Saint (1225). Canticle of the Creatures. In Regis Armstrong, J.A. Wayne Hellman, and William Short (eds.). *Francis of Assisi: Early Documents* (pp. 113–114). London-Manila, 1999.

Gaussin, M., Hu, G., Abolghasem, S., Basu, S., Shankar, M., & Bidanda, B. (2013). Assessing the Environmental Footprint of Manufactured Products: A Survey of Current Literature. *International Journal of Production Economics, 146*, 515–523.

Genazzani, A.R., Fidecicchi, T., Arduini, D., Giannini, A., & Simoncini, T. (2023). Hormonal and Natural Contraceptives: A Review on Efficacy and Risks of Different Methods for an Informed Choice. *Gynecological Endocrinology, 39*, 2247093.

Gómez-Heras, J.M. (2012). *Bioética y ecología. Los valores de la naturaleza como norma moral*. Madrid: Síntesis.
Gosling, D.L. (2001). *Religion and Ecology in India and Southeast Asia*. London: Routledge.
Green, K. (1996). Two Distinctions in Environmental Goodness. *Environmental Values, 5*, 31–46.
Gudynas, E. (2014). *Derechos de la Naturaleza y políticas ambientales*. La Paz: Plural Editores.
Guha, R., & Martinez-Alier, J. (1997). *Varieties of Environmentalism: Essays North and South*. London: Earthscan Publications.
Guillebaud, J., & Moore, P. (2009). Population Matters: Voluntary Contraception for Environmental Sustainability. In R.S. White (Ed.), *Creation in Crisis. Christian Perspectives on Sustainability* (pp. 84–101). London: SPCK.
Habel, N.C. (2011). *The Birth, the Curse and the Greening of Earth: An Ecological Reading of Genesis 1-11*. Sheffield: Sheffield Phoenix Press.
Hamilton, L., & McMillan, L. (Eds.) (2004). *Guidelines for Planning and Managing Mountain Protected Area*. Gland (S), Cambridge (UK): IUCN.
Hardin, G. (1968). The Tragedy of the Commons. *Science, 162*, 1248.
Himalayan Academy (2007). *What Is Hinduism?* Kapaa, Hawaii: Himalayan Academy.
Hornsey, M.J., Harris, E.A., & Fielding, K.S. (2018). Relationships among Conspiratorial Beliefs, Conservatism and Climate Scepticism across Nations. *Nature Climate Change, 8*, 614–620.
Houghton, J. (2006). Stewardship for the Twenty-first Century. In R.J. Berry (Ed.), *Environmental Stewardship* (pp. 315–317). London: T&T Clark International.
Intergovernmental Panel on Climate Change (IPCC) (2021). *Climate Change 2021: The Physical Science Basis. Contribution of Working Group I to the Sixth Assessment Report of the Intergovernmental Panel on Climate Change*. Cambridge, UK and New York: Cambridge University Press.
Intergovernmental Panel on Climate Change (IPCC) (2022). *Climate Change 2022: Impacts, Adaptation and Vulnerability. Working Group II Contribution to the IPCC Sixth Assessment Report*. Cambridge, UK and New York: Cambridge University Press.
International Theological Commission (2009). *In Search of a Universal Ethic: A New Look at the Natural Law*. Vatican: www.vatican.va/roman_curia/congregations/cfaith/cti_documents/rc_con_cfaith_doc_20090520_legge-naturale_en.html, accessed April 2021.
Ivanova, D., Vita, G., Steen-Olsen, K., Stadler, K., Melo, P.C., Wood, R., & Hertwich, E.G. (2017). Mapping the Carbon Footprint of EU Regions. *Environmental Research Letters, 12* (5):1–13.
Jagannathan, S. (1984). *Hinduism. An Introduction*. Mumbai: Vakils, Feffer and Simons.
Katz, E. (2002). The Call of the Wild. In: Scmidtz, D. & Willot, E. *Environmental Ethics. What Really Matters, What Really Works*. New York: Oxford University Press.
John Paul II (1987). *Solicitudo Rei Socialis*. Vatican. www.vatican.va/content/john-paul-ii/en/encyclicals/documents/hf_jp-ii_enc_30121987_sollicitudo-rei-socialis.html.
John Paul II (1990). *Message for the Celebration of the World Day of Peace*. Vatican: www.vatican.va/content/john-paul-ii/en/messages/peace/documents/hf_jp-ii_mes_19891208_xxiii-world-day-for-peace.html.

John Paul II (1991). *Centessimus annus*. Vatican: www.vatican.va/content/john-paul-ii/en/encyclicals/documents/hf_jp-ii_enc_01051991_centesimus-annus.html.
Johnson, L.E. (1992). Toward the Moral Considerability of Species and Ecosystems. *Environmental Ethics, 14* (2):145–157.
Jonas, H. (1995). *El principio de responsabilidad*. Barcelona: Herder.
Keller, D.R. (2010). *Environmental Ethics: The Big Questions*. Hoboken, NJ: Wiley-Blackwell.
Kellert, S.R. (2005). *Building for Life: Designing and Understanding the Human–Nature Connection*. Washington, DC: Island Press.
Kellert, S.R., & Wilson, E. (1995). *The Biophilia Hypothesis*. Washington, DC: Island Press.
Key, T.J., Appleby, P.N., & Rosell, M.S. (2006). Health Effects of Vegetarian and Vegan Diets. *Proceedings of the Nutrition Society, 65*, 35–41.
Kim, Y. (2017). St. Francis of Assisi and Ecological Conversion. *Catholic Theology and Thought, 78*, 47–89.
Kingsbury, D.V., Kramarz, T., & Jacques, K. (2019). Populism or Petrostate?: The Afterlives of Ecuador's Yasuní-ITT Initiative. *Society & Natural Resources, 32*, 530–547.
Koenig-Bricker, W. (2009). *Ten Commandments for the Environment: Pope Benedict XVI Speaks Out for Creation and Justice*. Notre Dame, IN: Ave Maria Press.
Krakauer, J. (1998). *Into the Wild*. London: Macmillan.
Leopold, A. (1949). *A Sand County Almanac. And Sketches Here and There*. New York: Oxford University Press.
Light, A. (2006). Ecological Restoration and the Culture of Nature: A Pragmatic Perspective. In Light, A. & Rolston, H. *Environmental Ethics. An Anthology*. Malden, MA: Blackwell Publishing.
Lindsey, J. (Ed.) (2012). *The Environment by Pope Benedict XVI*. Huntington, IN: Our Sunday Visitor.
Liu, K., & Lin, B. (2019). Research on Influencing Factors of Environmental Pollution in China: A Spatial Econometric Analysis. *Journal of Cleaner Production, 206*, 356–364.
Lorbiecki, M. (2014). *Following St. Francis: John Paul II's Call for Ecological Action*. New York: Rizzoli Ex Libris.
Lorentzen, L.A. (2001). *Etica Ambiental*. Puebla: Universidad Iberoamericana Puebla.
Louv, R. (2008). *Last Child in the Woods: Saving Our Children from Nature-Deficit Disorder*. Chapel Hill, NC: Algonquin Books.
Lovelock, J. (1979). *Gaia: A New Look at Life on Earth*. Oxford: Oxford University Press.
Lovelock, J. (2006). *The Revenge of Gaia: Why the Earth Is Fighting Back and How We Can Still Save Humanity*. New York: Penguin.
Maathai, W.W. (2006). *Unbowed: One Woman's Story*. (pp. 137–138). Nairobi: University of Nairobi.
Mackenzie, A., Ball, A.S., & Virdee, S.R. (2001). *Ecology*. Abingdon: Taylor & Francis.
Mallarach, J.M. (2012). *Spiritual Values of Protected Areas of Europe. Workshop Proceedings*. Bonn: Federal Agency for Nature Conservation.
Mallarach, J.M., Corcó, J., & Papayannis, T. (2014). Christian Monastic Communities Living in Harmony with the Environment: An Overview of Positive Trends and Best Practices. *Studia monastica, 56*, 353–391.
Marcos, A. (2001). *Ética Ambiental. Valladolid. Serie Filosofia. Colección "Acceso al saber"*. Valladolid: Universidad de Valladolid.

Marcos, A., & Pérez Marcos, M. (2018). *Meditación de la naturaleza humana*. Madrid: Biblioteca de Autores Cristianos.

Martinez de Anguita, P. (2002). *La tierra prometida*. Pamplona: Eunsa.

Masaki, K., & Tshering, J. (2021). Exploring the Origins of Bhutan's Gross National Happiness. *Journal of South Asian development, 16*, 273–292.

McDonagh, S. (1986). *To Care for the Earth: A Call to a New Theology*. London: Geoffrey Chapman.

Meadows, D.H., Randers, J., & Meadows, D. (1972). *The Limits to Growth*. New York: Universe books

Merino, J.A. (2008). *Francisco de Asis y la ecología*. Madrid: PPC.

Morales, J. (2001). *El Islam*. Madrid: Rialp.

Muir, J. (1912). *The Yosemite*. Washington, DC: Library of Congress Call Number F868. Y6 M9.

Myhre, G., Shindell, D., Bréon, F., Collins, W., Fuglestvedt, J., Huang, J., Koch, D., Lamarque, J., Lee, D., & Mendoza, B. (2013). *Anthropogenic and Natural Radiative Forcing, Climate Change 2013: The Physical Science Basis. Contribution of Working Group I to the Fifth Assessment Report of the Intergovernmental Panel on Climate Change, 659–740*. Cambridge: Cambridge University Press.

Naess, A. (1973). The Shallow and the Deep, Long-range Ecology Movement. A summary. *Inquiry, 16*, 95–100.

Naidoo, R., Balmford, A., Costanza, R., Fisher, B., Green, R.E., Lehner, B., Malcolm, T., & Ricketts, T. (2008). Global Mapping of Ecosystem Services and Conservation Priorities. *Proceedings of the National Academy of Sciences, 105*, 9495–9500.

Narayan, R. (2003). Ecological Crisis and Hindu Religious Thought. In R. Narayan & J. Kumar (Eds.), *Ecology and Religion* (pp. 25–37). New Delhi: Deep&Deep Publications.

Nasr, S.H. (2003). Islam, the Contemporary Islamic world, and the Environmental Crisis. In R. Folz, F. Denny & A. Baharuddin (Eds.), *Islam and Ecology* (pp. 85–105). Cambridge, MA: Harvard University Press.

Nasr, S.H. (2006). La naturaleza en el chamanismo y las religiones indígenas. In A. López Tobajas & M. Tabuyo (Eds.), *La naturaleza y el espíritu* (p. 106). Palma de Mallorca: J.J. de Olañeta.

Norton, B.G. (1984). Environmental Ethics and Weak Anthropocentrism. *Environmental Ethics, 6*, 131–148.

Norton, B.G. (1992). Epistemology and Environmental Values. *The Monist, 75*, 208–226.

Oki, T., & Kanae, S. (2006). Global Hydrological Cycles and World Water Resources. *Science, 313*, 1068–1072.

Okin, G.S. (2017). Environmental Impacts of Food Consumption by Dogs and Cats. *PLoS One, 12*, e0181301.

Ozdemir, I. (2003). Towards and Understanding of Environmental Ethics from a Quranic Perspective. In R. Folz, F. Denny & A. Baharuddin (Eds.), *Islam and Ecology*. Cambridge, MA: Harvard University Press.

Palmer, C. (2006). Stewardship: A Case Study in Environmental Ethics. In R.J. Berry (Ed.), *Environmental Stewardship* (pp. 63–75). London: T&T Clark International.

Papa Francisco (2020). *Soñemos juntos. El camino a un mundo futuro mejor*. Barcelona: Penguin Random House.

Pasad, R.K. (2003). Ecology and Buddhism. In R. Narayan & J. Kumar (Eds.), *Ecology and Religion* (pp. 41–54). New Delhi: Deep&Deep Publications.

Pathak, H., Jain, N., Bhatia, A., Patel, J., & Aggarwal, P.K. (2010). Carbon Footprints of Indian Food Items. *Agriculture, Ecosystems & Environment, 139*, 66–73.

Petersson, T., Secondi, L., Magnani, A., Antonelli, M., Dembska, K., Valentini, R., Varotto, A., & Castaldi, S. (2021). A Multilevel Carbon and Water Footprint Dataset of Food Commodities. *Scientific Data, 8*, 127.

Pope Francis (2013a). *General Audience, June 5th.* Vatican: www.vatican.va/content/francesco/en/audiences/2013/documents/papa-francesco_20130605_udienza-generale.html.

Pope Francis (2013b). *Homily during the Eucharistic Celebration at the Beginning of His Pontificate.* Vatican: www.vatican.va/content/francesco/en/homilies/2013/documents/papa-francesco_20130319_omelia-inizio-pontificato.html.

Pope Francis (2015). *Encyclical Letter Laudato Si'. On Care for Our Common Home.* Vatican: www.vatican.va/content/francesco/en/encyclicals/documents/papa-francesco_20150524_enciclica-laudato-si.html.

Pope Francis (2020). *Post-Synodal Apostolic Exhortation 'Querida Amazonia'.* Vatican: www.vatican.va/content/francesco/es/apost_exhortations/documents/papa-francesco_esortazione-ap_20200202_querida-amazonia.html.

Regan, T. (1983). *The Case for Animal Rights*. Berkeley: University of California Press.

Rockström, J., Steffen, W., Noone, K., Persson, Å., Chapin III, F.S., Lambin, E., Lenton, T.M., Scheffer, M., Folke, C., & Schellnhuber, H.J. (2009). Planetary Boundaries: Exploring the Safe Operating Space for Humanity. *Ecology and Society, 14* (2), 32.

Rolston, H. (1988). *Environmental Ethics. Duties to and Values in the Natural World.* Philadelphia, PA: Temple University Press.

Sagoff, M. (1984). Animal Liberation and Environmental Ethics: Bad Marriage, Quick Divorce. *Osgoode Hall Law Journal, 22*, 297–207.

Sala, S., & Castellani, V. (2019). The Consumer Footprint: Monitoring Sustainable Development Goal 12 with Process-based Life Cycle Assessment. *Journal of Cleaner Production, 240* (9), 297, 1–16.

Schumacher, E.F. (1973). *Small Is Beautiful. Economics as if People Mattered.* New York: Harper and Row.

Schwartz, E. (2002). Mastery and Stewardship, Wonder and Connectedness: A Typology of Relations to Nature in the Jewish Text and Tradition. In H. Tirosh-Samuelson (Ed.), *Judaism and Ecology* (pp. 93–106). Cambridge, MA: Harvard University Press.

Scultz, W. (2002). Empathizing with Nature: The Effects of Perspective Taking on Concern for Environmental Issues. *Journal of Social Issues, 56*, 391–406.

Shiva, V. (1991). *The Violence of the Green Revolution: Third World Agriculture, Ecology and Politics*. London: Zed Books.

Simon, J.L. (1981). *The Ultimate Resource*. Princeton, NJ: Princeton University Press.

Singer, P. (1975). *Animal Liberation*. New York: Harper Collins.

Sorrell, R.D. (1988). *St. Francis of Assisi and Nature: Tradition and Innovation in Western Christian attitudes toward the environment* New York–Oxford: Oxford University Press.

Soule, M.E. (1985). What Is Conservation iology? *BioScience, 35*, 727–734.

Steffen, W., Richardson, K., Rockström, J., Cornell, S.E., Fetzer, I., Bennett, E.M., Biggs, R., Carpenter, S.R., De Vries, W., & De Wit, C.A. (2015). Planetary

Boundaries: Guiding Human Development on a Changing Planet. *Science, 347*, 1259855.
Stern, N. (2007). *Stern Review on the Economics of Climate Change. Executive Summary*. London: M Treasury.
Stickler, C.M., Nepstad, D.C., Coe, M.T., McGrath, D.G., Rodrigues, H.O., Walker, W.S., Soareas-Filho, B.S., & Davidson, E.A. (2009). The Potential Ecological Costs and Cobenefits of REDD: A Critical Review and Case Study from the Amazon Region. *Global Change Biology, 15*, 2803–2824.
Stone, C.D. (1972). Should Trees Have Standing–Toward Legal Rights for Natural Objects. *Southern California Law Review, 45*, 450–487.
Swetnam, T.W. (1993). Fire History and Climate-Change in Giant Sequoia Groves. *Science, 262*, 885–889.
Tal, A. (2006). *Speaking of Earth: Environmental Speeches that Moved the World*. Piscataway, NJ: Rutgers University Press.
Taylor, P.W. (1986). *Respect for Nature: A Theory of Environmental Ethics*. Princeton, NJ: Princeton University Press.
Thakur, V.K. (2003). Ecological Perception of Buddhism in India. In R. Narayan & J. Kumar (Eds.), *Ecology and Religion* (pp. 58–59). New Delhi: Deep&Deep Publications.
Thoreau, H.D. (1854). *Walden or Life in the Woods*. New York: Dover Publications.
Tirosh-Samuelson, H. (2002). Introduction. Judaism and the Natural World. In H. Tirosh-Samuelson (Ed.), *Judaism and Ecology* (pp. xxxiii–lxii). Cambridge, MA: Harvard University Press.
Turvey, J.L. (2016). *Germain Grisez's Natural Law and Creation Theology as a Framework for Reflection on Climate Change and the Ecological Crisis*. Doctoral dissertation. University of Liverpool.
Valera, L., & Marcos, A. (2014). Desarrollo humano sostenible: una visión aristotélica. *Isegoría*, 671–690.
Vancutsem, C., Achard, F., Pekel, J.-F., Vieilledent, G., Carboni, S., Simonetti, D., Gallego, J., Aragão, L.E.O.C., & Nasi, R. (2021). Long-Term (1990–2019) Monitoring of Forest Cover Changes in the Humid Tropics. *Science Advances, 7*, eabe1603.
Vesilind, P.J. (1993). The Middle East's Water. *National Geographic, 183*, 38–71.
Victor, P.A. (2020). Cents and Nonsense: A Critical Appraisal of the Monetary Valuation of Nature. *Ecosystem Services, 42*, 101076+.
Vilkka, L. (1997). *The Intrinsic Value of Nature*. Amsterdam: Rodopi.
Ward, P.D., & Brownlee, D. (2000). *Rare Earth: Why Complex Life Is Uncommon in the Universe*. New York: Copernicus Books.
Waskow, A.O. (2000). *Torah of the Earth: Exploring 4,000 Years of Ecology in Jewish Thought*. Woodstock: Jewish Lights Publishing.
White, L.T. (1967). The Historical Roots of Our Ecological Crisis. *Science, 155*, 1203–1207.
Wilson, E.O. (2016). *Half-earth: Our Planet's Fight for Life*: New York: WW Norton.
Wirzba, N. (2003). The Character of Creation. *The Paradise of God: Renewing Religion in an Ecological Age* (pp. 23–70). New York: Oxford University Press.
Wolfenson, J.D. (2003). Foreword. In M. Palmer & V. Finlay (Eds.), *Faith in Conservation. New Approaches to Religions and the Environment* (p. xi). Washington, DC: World Bank.

Woods, T.E. (2005). *How the Catholic Church built Western Civilization*. Washington, DC: Regnery Publications.
Wunder, S. (2005). *Payment for Ecosystem Services: Some Nuts and Bolts*. Bogor, Indonesia: CIFOR Occasional Papers 42.
Wuputra, K., Ku, C.-C., Wu, D.-C., Lin, Y.-C., Saito, S., & Yokoyama, K.K. (2020). Prevention of Tumor Risk Associated with the Reprogramming of Human Pluripotent Stem Cells. *Journal of Experimental and Clinical Cancer Research, 39*, 1–24.
Wyatt, J. (2012). Temas éticos en el comienzo y fin de la vida. In E. Chuvieco & D. Alexander (Eds.), *Ciencia y religión en el siglo XXI: recuperar el diálogo* (pp. 283–296). Madrid: Editorial Centro de Estudios Ramón Areces.
Zizioulas, J. (2013). Proprietors or Priests of Creation? In J. Chryssavgis & B.V. Foltz (Eds.), *Toward an Ecology of Transfiguration: Orthodox Christian Perspectives on Environment, Nature, and Creation* (pp. 163–172). New York: Fordham University Press.
Zizioulas, J. (2021). Orthodoxy and the Ecological Crisis. In J. Chryssavgis & N. Asproulis (Eds.), *Priests of Creation: John Zizioulas on Discerning an Ecological Ethos* (pp. 55–60). London: T&T Clark.

Index

FSC
www.fsc.org
MIX
Papier | Fördert
gute Waldnutzung
FSC® C083411

Kolibri 360 GmbH
Ferdinand-Jühlke-Straße 7
99095 Erfurt, Deutschland
produktsicherheit@kolibri360.de